Enquirybase Geography

3 World Issues

Stephen Scoffham

Colin Bridge

Terry Jewson

CONTENTS

1	People on the move	13	Spare resources?	25	Money to spend
2	Different peoples	14	Leisure facilities	26	Unequal shares
3	Lonely people	15	Tourist facilities	27	Points of view
4	Trade power	16	Microclimates	28	Different diets
5	Commodities	17	Disruption	29	Medical services
6	Multinationals	18	Trends	30	Learning for life
7	Different houses	19	Houses game	31	Pollution
8	Town or country?	20	Planning issues	32	Conservation
9	Desertification?	21	A new bypass?	33	Derelict land
10	Physical regions	22	Job opportunities	34	What matters most?
11	Industrial regions	23	Energy demand	35	Choices
12	Service regions	24	Forecasting	36	Tomorrow's world

SUMMARY SHEET

NUMBER	TITLE	DESCRIPTION	CONCEPT
1	People on the move	Analysis of pupils' birth places using desire lines	**Movement** Migration
2	Different people	Family names as a clue to immigration over history	
3	Lonely people	Subjective responses to different school structures	
4	Trade power	Trading game illustrating differences between nations	**Communication** Trade
5	Commodities	Effects of changes of demand on sugar-producing countries	
6	Multinationals	Survey of groceries produced by Unilever	
7	Different houses	Effect of climatic factors on traditional building styles	**Place** Landscapes
8	Town or country?	Classification of settlement patterns by map grid squares	
9	Desertification	Fieldwork study of 'arid areas' in front gardens	
10	Physical regions	Defining regions of Britain by relief, climate and geology	**Region** Regional analysis
11	Industrial regions	Field and mapwork study of industrial activity	
12	Regions	Comparison of local, regional and national organisations	
13	Spare resources?	Ways of putting school buildings to better use	**Resources** Leisure and tourism
14	Leisure facilities	Compatibility of different leisure activities	
15	Tourist facilities	Planning facilities to match holiday needs	
16	Microclimates	Analysis of microclimates in school building and grounds	**Pattern** Trends
17	Disruption	How we depend on other people in our daily lives	
18	Trends	Analysis of changing rolls in three sample schools	
19	Houses game	Designing a housing estate using template models	**Design** Planning issues
20	Planning issues	Role play of plans to redevelop a factory site	
21	A new bypass?	Simulation of proposals for a new by-pass	
22	Job opportunities	Changes in employment since the turn of the century	**Change** Technology
23	Energy demand	Line graph of daily electricity consumption	
24	Forecasting	Predicting short-term behaviour patterns	
25	Money to spend	Access to consumer goods in different parts of the world	**Structure** Differences in wealth
26	Unequal shares	Contrasts in housing worldwide	
27	Points of view	Analysis of how different people respond to the same environment	
28	Different diets	Survey of consumption of processed food in Britain and China	**Process** Living conditions
29	Medical services	Comparison of health care in Britain and Zimbabwe	
30	Learning for life	The North/South divide illustrated through illiteracy rates	
31	Pollution	Impact and duration of different types of pollution	**System** Pollution and conservation
32	Conservation	Questionnaire of personal behaviour patterns	
33	Derelict land	Possible uses for patches of derelict land	
34	What matters most?	Survey of problems that are most pressing in school	**Interaction** Choices
35	Choices	Impact of family size and wealth on the quality of life	
36	Tomorrow's world	Pupils vote for the things they most want in a perfect world	

GEOGRAPHICAL EXTENSION	NATIONAL CURRICULUM LINKS				
	AT1	AT2	AT3	AT4	AT5
Migrant workers, resettlement programmes and refugees Influence of different languages, customs and cultures Gypsies, nomads and tribal groups	• • •	•		• • •	
Historic trade routes. Colonialism. Terms of trade Commodity trade – tea, rubber, tin, coffee Importance of multinational companies in the world economy	• • •	• •		• • •	
Traditional and modern building styles worldwide Ribbon development, new estates and Green Belts Soil erosion and desertification – problems and solutions	• • •	• •	•		• •
World climatic regions, natural vegetation and agriculture Location of industry. Influence of markets and raw materials Centres of European and international administration	• • •	• • •	•	• •	
Under-used resources in Britain – canals, railways, human talent Games, sports, conservation groups and the environment History, development and impact of modern tourism	• • •	• • •			• • •
Coastal breezes, frost pockets and shelter belts Heatwaves, floods, blizzards, earthquakes and other disasters Recording, measuring and collecting geographical data	• • •		• •	• •	
Planning regulations, redevelopment schemes, new towns Contemporary building projects worldwide Different ways of solving traffic problems	• • •	• • •		• • •	
Simple, intermediate and advanced technology Seasonal demand for consumer goods. Cycles in the natural world Forecasting and advance planning	• • •	•		• • •	
GNP of different countries of the world Slums. Shanty towns. Regional differences in Britain Urban growth. Cities with over a million inhabitants	• • •	• • •		• • •	
Food preservation. Calorie intake. Malnutrition Life expectancy worldwide and the control of disease Schools, education and opportunities for training	• • •	• • •		• • •	
Pollution disasters and threatened environments Sustainable agriculture. Conservation schemes Waste disposal. Recycling. Land reclamation	• • •	•			• • •
World problems. Brandt and Brundtland Reports Living conditions worldwide International conflicts and the work of the United Nations	• • •	•		• •	• • •

INTRODUCTION

Enquirybase Geography is a series of three books designed to provide fieldwork and other practical activities for pupils at key stage three. Each book contains thirty-six pupil activity sheets and extensive notes and resources for the teacher. The activity sheets can be photographed for classroom use.

The geography national curriculum requires pupils to engage in investigations and enquiries. These are not intended to be optional extras but a fundamental part of the curriculum. Fieldwork is a particularly important element as it enables pupils to develop a range of skills that they will need in later life.

Enquirybase Geography meets the need for on site work with structured activities which can be carried out in the classroom, school building, school grounds and nearby streets. Pupils who follow the course will be introduced to a wide range of techniques, thereby reducing the need for visits to distant locations and disruption to the timetable. As an added advantage they will examine geographical aspects of environments with which they are already familiar.

All the activities involve data collection and the careful analysis of findings. Maps, diagrams, graphs, symbols, codes and annotated drawings are introduced in appropriate contexts. Discussion and the use of geographical vocabulary form a vital part of the programme, and specialist terms are introduced where appropriate. A variety of issues, questions and problems are also raised to give pupils experience in constructing and testing their own hypotheses, and to allow them to consider attitudes and values.

Pupils can complete the activities working either on their own or in groups. The teacher's notes contain suggestions for further investigations and homework. These studies all adopt the enquiry approach advocated by the national curriculum.

Structure

The books are structured by concept. There are three activities for each concept and twelve concepts in each book. This approach is readily compatible with the national curriculum strands and means that the material can be easily related to studies of places and themes.

There is also a concentric progression within the series. Book 1 looks at the school and its surroundings, Book 2 considers the neighbourhood and Book 3 takes a wider world perspective. The matrix below shows how this is achieved.

CONCEPT	BOOK 1	BOOK 2	BOOK 3
MOVEMENT	journeys	traffic	migration
COMMUNICATION	messages	transport	trade
PLACE	settlement	urban growth	landscapes
REGION	zones and areas	land use	regional analysis
RESOURCES	housing	industry	leisure and tourism
PATTERN	behaviour patterns	shopping	trends
DESIGN	building design	improving the environment	planning issues
CHANGE	neighbourhoods	population	technology
STRUCTURE	classification	hierarchies	differences in wealth
PROCESS	society	employment	living conditions
SYSTEM	eco-systems	services	pollution and conservation
INTERACTION	interdependence	people in the environment	choices

Teacher's notes

Each study is presented as a double-page spread with teacher's notes and resources on the left-hand side and pupils' activity sheet on the right. The notes describe the activity, place it in a broad geographical context and explore the main concepts involved. They also give clear guidance as to essential skills and preparation required and how the activity might be extended and developed. For ease of reference, a standard format has been adopted under the following headings.

1 Skills
Lists the main skills used in the activity. Provides links with record-keeping requirements.

2 Attitudes and values
Highlights the issues raised in the activity and suggests ways in which they might be explored or discussed.

3 Lesson preparation
Lists main teaching points, special equipment and vocabulary required in the activity. Helps to ensure a high degree of success in completion.

4 Local enquiry/homework
Extends the activity through fieldwork and surveys in the locality. Includes suggestions for homework and individual research.

5 Extended investigation
Ideas for further work that extends the activity into regional, national and world studies. Aids links with textbooks and special topics.

6 Problem
Open-ended questions for pupils to explore in a geographical way using ideas and techniques from the activity.

7 Data resources
Provides a range of information which can be used for further discussion, analysis, research or enquiry. Also provides addresses and references as necessary.

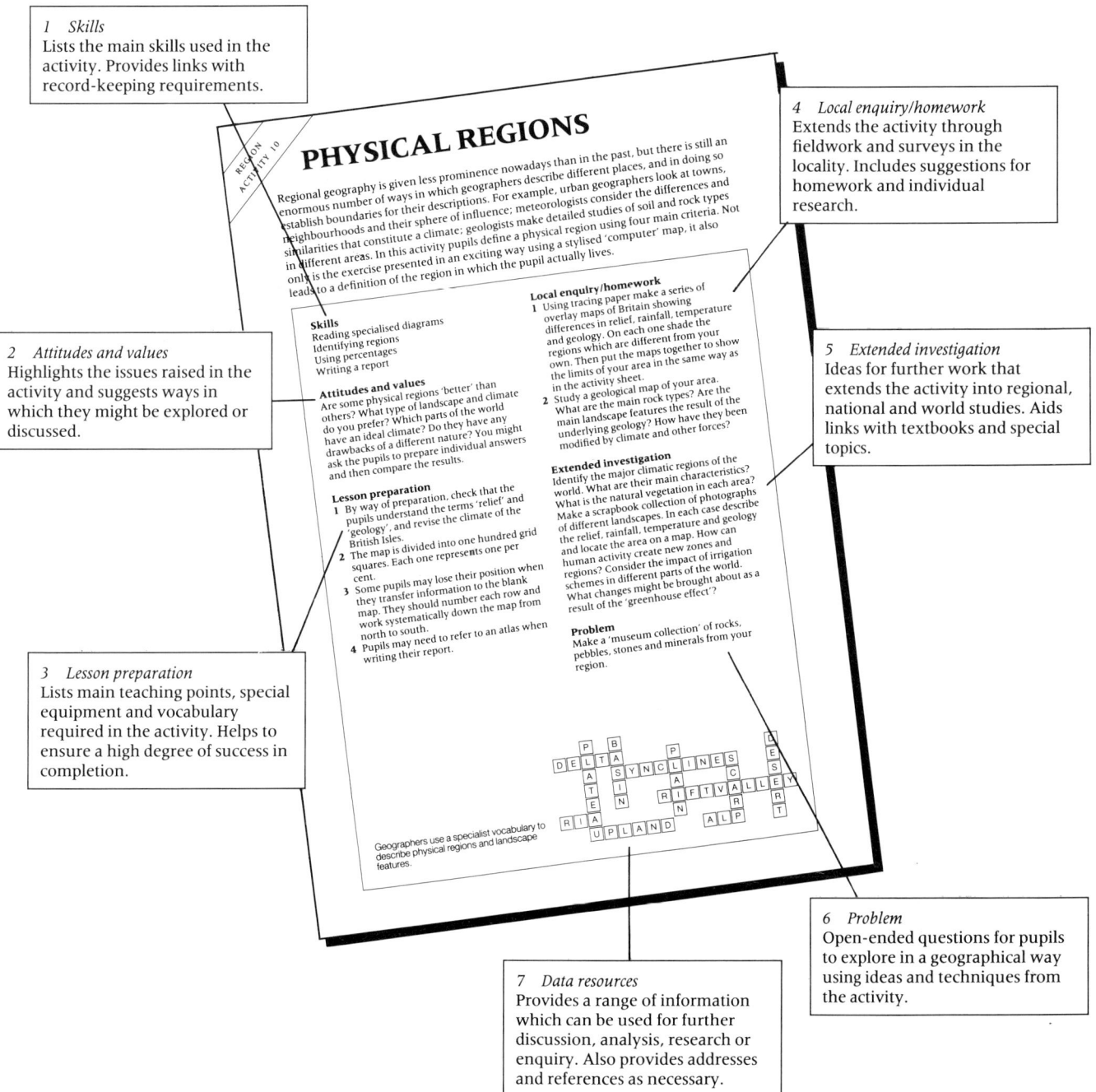

Activity sheets

The activity sheets are also standardised. The title encapsulates the theme which is then expanded in a succinct geographical statement. The activity itself is broken down into a series of clearly numbered stages, starting with simple data collection and leading to analysis and presentation of findings. The last question is usually open-ended and, if possible, suggests further practical work in the immediate environment.

It is important that pupils work through each activity in a systematic way, following the numbers on the page. They may also wish to present their results on separate sheets of paper, perhaps using maps and diagrams on a larger scale. Excellent results can be achieved for exhibitions and displays.

	NAME OF PUPIL	CONCEPT			SKILLS				EFFORT				NOTES
	Marking Scheme − needs help √ satisfactory + doing well or graded A, B, C, D, E, F, G or marks out of ten	understanding of concept	attitudes and values	problem work	collecting information	organising information	analysing information	mapwork	extended investigation	homework	quality of presentation	TOTAL (out of 10)	
Movement	1 People on the move												
	2 Different peoples												
	3 Lonely people												
Communication	4 Trade power												
	5 Commodities												
	6 Multinationals												
Place	7 Different houses												
	8 Town or country?												
	9 Desertification												
Region	10 Physical regions												
	11 Industrial regions												
	12 Service regions												
Resources	13 Spare resources?												
	14 Leisure facilities												
	15 Tourist facilities												
Pattern	16 Microclimates												
	17 Disruption												
	18 Trends												
Design	19 Houses game												
	20 Planning issues												
	21 A new bypass?												
Change	22 Job opportunities												
	23 Energy demand												
	24 Forecasting												
Structure	25 Money to spend												
	26 Unequal shares												
	27 Points of view												
Process	28 Different diets												
	29 Medical services												
	30 Learning for life												
System	31 Pollution												
	32 Conservation												
	33 Derelict land												
Interaction	34 What matters most?												
	35 Choices												
	36 Tomorrow's world												

CLASS	TITLE OF ACTIVITY																																				
NAME OF PUPIL	1 People on the move	2 Different peoples	3 Lonely people	4 Trade power	5 Commodities	6 Multinationals	7 Different houses	8 Town or country?	9 Desertification	10 Physical regions	11 Industrial regions	12 Service regions	13 Spare resources?	14 Leisure facilities	15 Tourist facilities	16 Microclimates	17 Disruption	18 Trends	19 Houses game	20 Planning issues	21 A new bypass?	22 Job opportunities	23 Electricity demand	24 Forecasting	25 Money to spend	26 Unequal shares	27 Points of view	28 Different diets	29 Medical services	30 Learning for life	31 Pollution	32 Conservation	33 Derelict land	34 What matters most?	35 Choices	36 Tomorrow's world

MOVEMENT ACTIVITY 1

PEOPLE ON THE MOVE

The movement of people outwards from Central Asia has been one of the key dynamics in human history. In Britain, for example, wave after wave of immigrants arrived from the east seeking land, trade, conquest and exploration. The strife and conflict that ensued were the birth pangs of the modern nation. In this activity pupils consider movement in contemporary society. They plot the place of birth of ten sample children on a map of Britain and make a similar survey in their own school. This illustrates the fact that many families are highly mobile, and challenges the pupils to consider how local cultures and traditions can best be preserved.

Skills
Using an atlas
Calculating a distance using a scale
Using a desire-line diagram
Conducting a survey

Attitudes and values
The activity raises complex and emotive issues. For example, should people have the right to move freely from one country to another, or is there a need for restrictions? You might consider current British immigration policy. Is it consistent, fair and just, or could it be improved?

Lesson preparation
1 Remind the pupils how to calculate distances using a scale.
2 It is helpful to identify the places shown on the map before completing the table. The simplest is to write the initial letter of the town next to each dot.
3 Pupils will need an atlas to find the place of birth of people in their survey and a pair of compasses to draw distance rings on the map.
4 Question 5 should prompt discussion about the economic structure of the locality.

Local enquiry/homework
1 Make a survey of people in your street to find out how long they have lived there. Draw a map to show your results. Add notes about what brought you to the district and the places your family has lived in the past.
2 Can you find any evidence in your area that people are regularly on the move? Make a collection of clues. Find out about removal firms, estate agents and other industries and services which help people to move home.

Extended investigation
In what parts of the world are people currently moving about in large numbers? Using newspaper articles make a sample study of refugees, migrant workers or resettlement programmes. What causes people to move? Are natural or human factors more important? Find out about the social problems created by mass movements. How do international organisations try to provide help, and are they effective?

Problem
Make up a short introductory guide to help a newcomer fit into your community as quickly and easily as possible.

Principal refugee and displaced person populations in 1989.

PEOPLE ON THE MOVE

Many people are newcomers to the area where they live.

1 Here is a list of pupils in a Manchester school. Mark their place of birth on the map, using an atlas to help you. Then tick the correct column in the table to show how far they have moved.

NAME	PLACE OF BIRTH	0–50	50–100	100–200	200–300	Over 300
Tony O'Neil	Belfast					
Richard Pope	Manchester					
Wendy Pearson	Leeds					
Murli Mulchandani	Bombay, India					
Saiqa Balagh	Manchester					
Anthony Harrison	Glasgow					
Nabila Hussain	Sheffield					
Fiona Donovan	Sheffield					
Tracey Adams	Birmingham					
Paul Thrower	Leeds					
TOTAL						

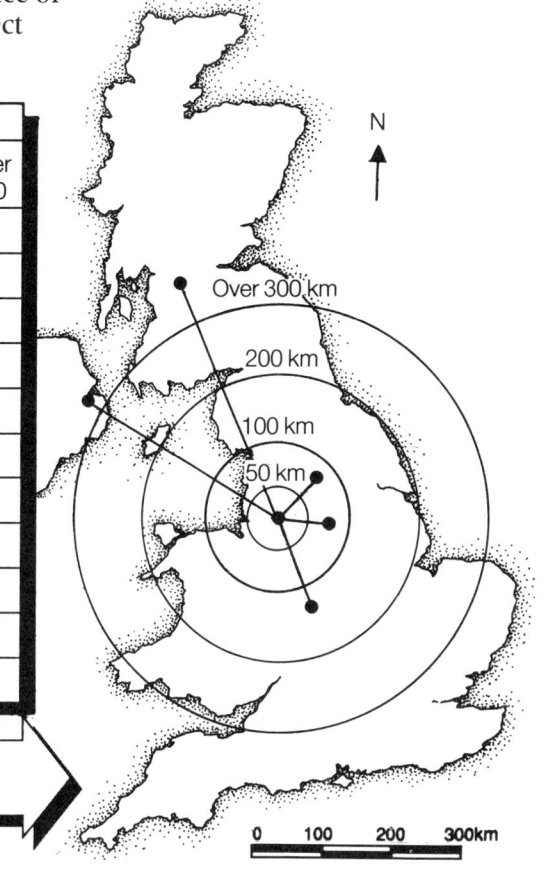

2 Colour the ring on the 'radial chart' which shows where most pupils come from.

3 Make a similar survey of ten pupils in your class.

NAME	PLACE OF BIRTH	0–50	50–100	100–200	200–300	Over 300
TOTAL						

4 Show your results on a map. Begin by marking your own town and the distance rings using compasses. Then mark the places people come from. Finally colour the ring with the highest total.

5 Is there any evidence that people are moving into or out of your area? Can you think of any reasons for this? Write a short report.

© Scoffham, Bridge, Jewson, 1991. Macmillan. *Enquirybase Geography, Book 3*   ACTIVITY 1

MOVEMENT ACTIVITY 2

DIFFERENT PEOPLES

Two thousand years ago Britain was inhabited by a range of people organised in small tribal groups. Gradually these have been assimilated and welded into a single nation with a distinct politial, cultural and social unity. In this activity pupils extract information from a timeline to discover how family names indicate the different ethnic background of different sections of the population. The exercise suggests that the invasions and cultural differences of the past have now lost their significance. By implication the differences which are still acute in contemporary society will also be softened by the process of assimilation.

Skills
Reading a timeline
Identifying ethnic groups
Making a bar graph
Conducting a survey

Attitudes and values
Some people set great store by their ethnic and cultural identity, while others mix more freely. You could discuss the advantages and disadvantages of a multi-cultural society. How have different groups contributed to modern Britain? Are they all given equal rights?

Lesson preparation
1 A study of common English words illustrates how language reflects the cultural forces that have shaped modern Britain and serves as a good introduction to the activity.
2 To complete the table, pupils will need to refer to the timeline and datafiles. Precise datas have been omitted deliberately and there are alternative dates for both Celtic and French immigration.
3 It is unlikely that pupils will be able to identify the origin of all the names in their class. However, they should be able to find at least some clear examples.

Local enquiry/homework
1 Make a collection of the names of local places, streets, businesses and shops. How many of them provide clues about the different groups of people who have lived in the area? Analyse your findings using a datafile, and say which influences seem most important.
2 Using a street directory, parish register or evidence from local gravestones, compile a list of names that are traditional to your area. What do they tell you about the history of the district?

Extended investigation
Make a study of common English words, using a dictionary. For each one write down its derivation or origin. Is there any other evidence of the influence of different cultures on our everyday life? How have the great empires (Roman, Islamic, Chinese) affected us? What special skills have people brought to this country from abroad, e.g. weavers from the Low Countries? Consider the impact of trade and communications. Has it been detrimental or beneficial?

Problem
What are the six most common names in your locality? Can you explain their derivation?

Until the 1980s more people were leaving Britain than arriving from abroad. Now the numbers are more or less equal.

Source: *OPCS Spotlight 4*, HMSO

DIFFERENT PEOPLES

Family names can help to identify different groups of people.

1 Look carefully at the datafile and timeline. They tell you when different groups of people arrived to live in Britain.

CELTIC NAMES	
Clue	Begin with Mac, Mc or O'
Examples	MacGregor, O'Brien
Also	Jones, Owen

ANGLO-SAXON NAMES	
Clue	Names of trades, colours or trees
Examples	Smith, Baker, Brown, Alder

SCANDINAVIAN NAMES	
Clue	Often end in -son or -sen
Examples	Ericson, Neilson, Olsen

FRENCH NAMES	
Clue	Begin with Le or De (D') or contain French words
Examples	Lemar, Disney, Petit, Beaumont

INDIAN/PAKISTANI NAMES	
Clue	Very varied, but often contain a silent h
Examples	Singh, Khan
Also	Patel, Kaur

CHINESE NAMES	
Clue	Often end in -ang, -ong or -ing
Examples	Tang, Wong

Timeline:
- Celts — 100 BC
- Romans — 0–400
- Anglo-Saxons — 400–600
- Vikings (Scandinavian) — 800 AD
- Normans (French) — 1000
- Huguenots (French) — 1500
- Irish (Celts) — 1800
- Indians/Pakistanis — 1900
- Chinese — 2000

2 Complete the table below by saying what group each name belongs to and when it might date from.

FAMILY NAME	GROUP	POSSIBLE DATE	FAMILY NAME	GROUP	POSSIBLE DATE
Smith			Larsen		
Patel			O'Donovan		
Miller			Jones		
Ash			Crabtree		
Khan			Whitehead		
Tang			Cooper		
Owen			Jansen		
Lemar			Petit		
Beer			Brown		
Palmer			Macmillan		

3 Draw a block graph to show the origin of the different names in the list.

NAMES	NUMBER IDENTIFIED
Celtic	
Anglo-Saxon	
Scandinavian	
French	
Indian/Pakistani	
Chinese	

4 Make a similar survey of names in your class or school.

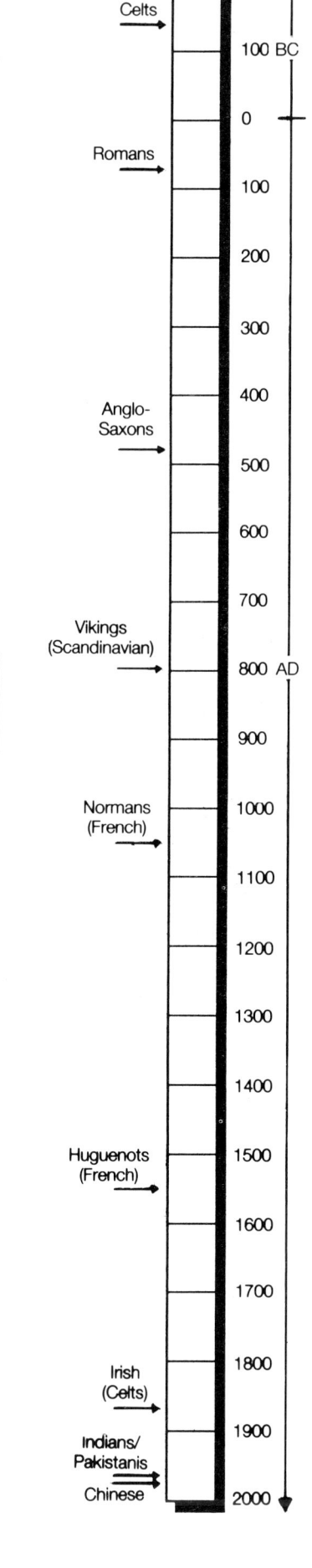

© Scoffham, Bridge, Jewson, 1991. Macmillan. *Enquirybase Geography, Book 3*

ACTIVITY 2

MOVEMENT ACTIVITY 3

LONELY PEOPLE

Although the movement and mixing of people may bring great benefits, it can be a painful process for those who are actually involved in it. Many immigrants feel rootless and isolated – strangers in the society which they have just joined, yet alienated from their country of origin. Pupils too may experience similar feelings of ennui as they move through the educational system, particularly when they arrive in a new school for the first time. This activity examines how the organisational structure of a large secondary school may create isolated minorities who feel daunted and threatened.

Skills
Constructing a timeline
Tabulating data
Comparing administrative systems
Eliciting opinions

Attitudes and values
The exercise raises important psychologial questions. For example, how do environmental factors contribute to or undermine a person's well-being? What factors are important in developing self-image and esteem? Are people limited or constrained by their expectations?

Lesson preparation
1 A brief discussion about the different types of secondary school, both past and present, helps to introduce the activity.
2 See that pupils understand how to complete the profile. For each school they must colour the positive or negative box, not both. All the information required is given in the timetables.
3 The words in the survey at the bottom of the sheet are arranged in pairs. They tend to be mutually exclusive. Explain this, and see that pupils understand terms such as 'anonymous' and 'valued'.

Local enquiry/homework
1 Make a study of your own locality. List what you consider to be the positive and negative features. Discuss this with others. Do they agree with you? Make a comparison between your area and another nearby place. What features seem most important in making people feel less lonely and isolated?
2 Make up a questionnaire to find out what local people think about their surroundings. Write a short report describing your findings.

Extended investigation
Make a study of a group of people, such as nomads or gypsies, who are regularly on the move. What are the main problems that they experience? How do these groups retain their identity? What are the special features that give them strength? Find out about tribal groups which are currently threatened with extinction. Consider the case of the Inuit or Aborigines. In what way are modern governments trying to cater for their needs?

Problem
Make up a diagnostic questionnaire to find out which people in your class have the qualities to live in isolated places.

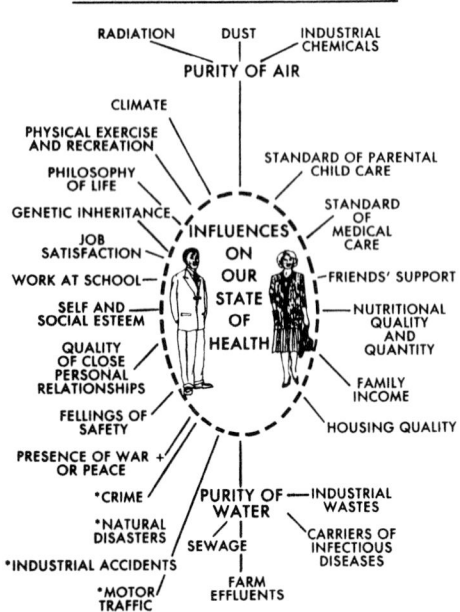

Source: *World Studies Journal*, Vol. 7, No. 1, 1988

LONELY PEOPLE

People who are always on the move tend to feel **lonely and isolated**.

1 Look at the table below. It gives the timetable for a typical day at (a) Surbiton Grammar School thirty years ago, (b) Danecourt Comprehensive School nowadays.

2 Add a timetable for a typical day at your own school, listing the lessons and the places where they happen.

SURBITON GRAMMAR (600 pupils)	DANECOURT COMPREHENSIVE (1600 pupils)	
Registration in form room at own desk	Registration in form room	
Assembly – whole school in hall	Assembly in form room	
English in form room	English in room D4	
Maths in form room	Maths in maths room	
Break in playground	Walk to other school site	
French in form room	Physics in science lab	
History in form room	Physics in science lab	
First sitting lunch at set table	Self-service lunch in cafeteria	
Chemistry in science lab	Geography in geography room	
Chemistry in science lab	Social studies in room P9	
Chemistry in science lab	Social studies in room P9	

3 Build up a profile for each school by colouring *one* box against each pair of words in the chart.

Your school	Danecourt	Surbiton	POSITIVE FEATURES (can build security, self-esteem and independence)	NEGATIVE FEATURES (can make people feel lonely and isolated)	Surbiton	Danecourt	Your school
			Less than 1000 pupils	More than 1000 pupils			
			Whole school meets regularly	School rarely meets together			
			Pupils recognise each other	Pupils remain strangers			
			Pupils mostly stay in form room	Pupils move around a lot			
			Pupils have own desks	Pupils have lockers			
			Lunch at set tables	Self-service lunch			
			TOTAL OF POSITIVE FEATURES	TOTAL OF NEGATIVE FEATURES			

4 Which school has the most positive features?

_____

5 Which school has the most negative features?

_____

6 Now make a survey of five pupils in your class to find out how they feel about school. Put a tick or a cross against each word.

7 Which words have the highest scores?

_____

NAME	secure	worried	safe	threatened	friendly	hostile	valued	anonymous	independent	controlled
TOTAL										

8 Explain what features of your school are important in creating its character.

© Scoffham, Bridge, Jewson, 1991. Macmillan. *Enquirybase Geography, Book 3*

TRADE POWER

COMMUNICATION ACTIVITY 4

Trade is a fundamental factor in the growth and development of many towns and communities. From the earliest times people have been keen to exchange their surplus crops and products. The voyages of exploration of the sixteenth century were prompted by the prospect of wealth and riches from overseas. Nowadays the world is linked by a global trading economy in which Britain especially depends on other nations for its goods and materials. This activity introduces the idea of trade and shows that some countries have much more power than others. Pupils may feel that the game is rather arbitrary, but then so is the accident of birth which gives some people so many more opportunities than others.

Skills
Playing a geographical game
Identifying countries by outline shape
Making comparisons
Analysing causes and explanations

Attitudes and values
Current trading patterns do not promote equality. Some of the poorest countries in the world, such as Tanzania and Bangladesh, are selling their products at a loss. Is this right? How could the situation be rectified? What action can you take as an individual?

Lesson preparation
1 The activity has been designed as a class simulation. Smaller groups will need to choose from a smaller range of numbers when playing the first round of bingo.
2 The game continues in Question 4 by using the export values at random, as in bingo. The values can be either read out from the list below, or drawn out of a hat. It is important that pupils accept that the game is 'fair' within its own terms.
3 The values of the exports for each country are given in millions of US dollars and are based on figures in the 1988 *Europa Handbook*.

aircraft	15,000	machinery	11,000
cars	14,000	medicines	2,500
chemicals	22,000	paper	1,000
clothes	2,000	petrol	14,000
cobalt	3	sacks and rope	300
coffee	2,500	scientific	
copper	250	instruments	3,000
electrical		shoes	1,500
goods	18,000	sugar	400
fish	80	tea	30
fruit	3,000	textiles	1,000
hides	60	tobacco	1
iron and steel	5,000	wheat	3,000
iron ore	1,500	wood	300
jute	100	vehicles	7,000
lead	1	zinc	7

4 To complete the table in Question 5 it is useful to collate information about how much each country has earned on the board.
5 The aim of the game is to give an accurate impression of the trading base of a number of different countries in the First and Third Worlds. Question 7 should provoke a good deal of discussion.

Local enquiry/homework
1 Make a survey of goods and materials in your own home or classroom which have come from abroad. Which countries do we seem to trade with most? Can you explain why?
2 Find out how trade has influenced your own town or region. Describe the main activities at different times in the past. Using maps and diagrams, make a sample study of one selected trade or industry.

Extended investigation
Find out about historic trade routes such as the Silk Road between Europe and China. Draw a map showing different routes. Make a sample study of a colonial trading company such as the Hudson Bay or East India Company. What did they trade in, and how was the trade conducted? What were the 'trade winds'? Find out about 'free trade' and 'protectionism'. Explain the differences, using examples. What societies and organisations are designed to promote trade? Are there any examples in your own area or locality?

Problem
Make up your own trading game.

TRADE POWER

Some countries have much more trade power than others.

1 Choose any five numbers between one and forty and write them in the top row of boxes opposite.

GAME 1					
GAME 2					

2 Now play a round of bingo. Your teacher will read out the winning numbers. Circle them as they are called and shout 'trade' when you complete a line.

3 Find out which country you represent from the table below. Colour its outline and look at the list of exports. In a moment you will be told their values.

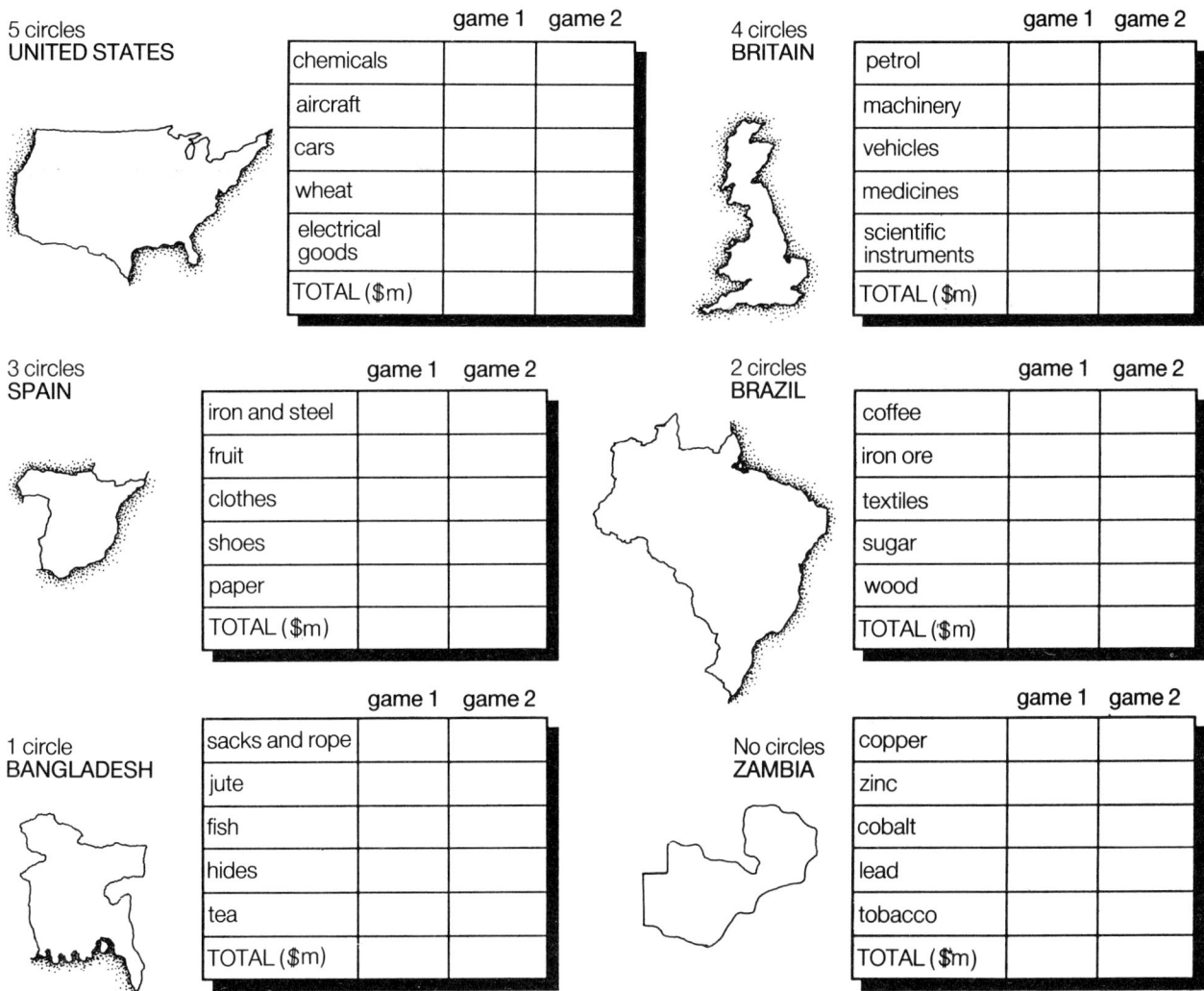

5 circles
UNITED STATES

	game 1	game 2
chemicals		
aircraft		
cars		
wheat		
electrical goods		
TOTAL ($m)		

4 circles
BRITAIN

	game 1	game 2
petrol		
machinery		
vehicles		
medicines		
scientific instruments		
TOTAL ($m)		

3 circles
SPAIN

	game 1	game 2
iron and steel		
fruit		
clothes		
shoes		
paper		
TOTAL ($m)		

2 circles
BRAZIL

	game 1	game 2
coffee		
iron ore		
textiles		
sugar		
wood		
TOTAL ($m)		

1 circle
BANGLADESH

	game 1	game 2
sacks and rope		
jute		
fish		
hides		
tea		
TOTAL ($m)		

No circles
ZAMBIA

	game 1	game 2
copper		
zinc		
cobalt		
lead		
tobacco		
TOTAL ($m)		

4 Now continue the game. Write down the value of your country's exports as they are read out. Shout 'trade' as soon as you have all five values. This stops the game.

5 Add up the total value of products traded by your country and complete the table below by class discussion.

	game 1	game 2
Country which sold the most products		
Country which earned the most money		

6 Play another game to find out what happens in a second trading year.

7 What causes differences in trade power? Consider each of the following statements in turn, and say if you think they are true or false. Use examples in your answer.

(a) Some countries have more natural resources than others.
(b) Some countries are larger than others.
(c) Differences in climate make it harder to live in some parts of the world.
(d) Some products are more valuable (scarcer) than others.
(e) Some countries have to sell their products at low prices in order to survive.

© Scoffham, Bridge, Jewson, 1991. Macmillan. *Enquirybase Geography, Book 3*   ACTIVITY 4

COMMUNICATION ACTIVITY 5

COMMODITIES

One of the consequences of the Industrial Revolution and the expansion of European empires was the growth of a world economy. From the eighteenth century onwards, food and agricultural production began to be dominated by European needs. Plantations were established in many parts of the world. These provided cash crops such as tea, coffee, rubber and sugar for the world market. This same pattern of trade still persists today. It is enforced by a system of economic dependence which has proved particularly pernicious in the last fifteen years as commodity prices have tumbled. In this activity pupils examine their own pattern of sugar consumption and see how it affects other countries.

Skills
Using a code
Sorting into sets
Making a bar graph
Making predictions

Attitudes and values
Many countries in Africa and South America depend on a single product for the bulk of their export income. You might discuss how this situation evolved and the problems that it is causing nowadays. Sugar presents a particularly interesting case study as there are alternative sources of supply from beet, maize and artificial sweeteners. In these circumstances, do we have a responsibility to continue importing cane sugar from the Tropics?

Lesson preparation
1 The figures in the table show the quotas for preferential imports to the EC under the Sugar Protocol of the Lome Convention for 1985.
2 Before beginning, pupils should identify the countries listed on the table in an atlas and discuss the differences between cane and beet sugar.
3 Question 7 is designed to promote discussion. A cut of 40,000 tonnes per country would put Belize out of business completely.

Local enquiry/homework
1 Make a survey of the food in your home. List the contents of the refrigerator, food cupboard or larder. How many of the items contain sugar, glucose or artificial sweeteners? Which are sugar-free or designed as a 'diet' food?
2 Make a collection of advertisements which promote 'sugar-free' products. What image are they trying to present?

Extended investigation
Make a list of other commodities which are supplied by the Third World – rubber, tea, coffee, tin, copper, oil and so on. Investigate one of them. How is it produced? When did the trade begin ? What are the living conditions of the workers? Do the producers get a fair price? Use maps, diagrams and pictures to illustrate your study. To what extent does Britain depend on the international commodity trade? Try to identify the origin of goods and materials in your school or home.

Problem
Devise a day's menu for yourself which would involve no sugar consumption of any kind.

Most cash crops have been transferred from one part of the world to another.

Source: *Contemporary Issues in Geography and Education*, Vol. 1, No. 3

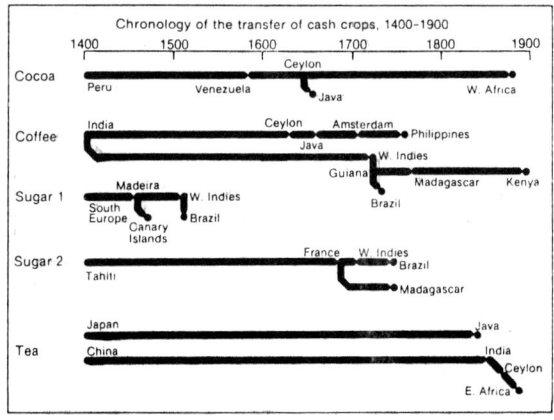

Source: *Third World Atlas*, Open University, 1983

COMMODITIES

In some countries people depend on one commodity for their livelihood.

TYPE OF SWEETENER	CODE
Sugar	S
Glucose	G
Artificial sweeteners	A

1 Find out from the codes what sweetener each of the foods in the list contains.

2 Look at the list of foods again and tick the ones you eat regularly.

3 Write down the foods you eat regularly in the correct column of the table.

SUGAR	MORE THAN ONE SWEETENER

FOOD	CODE	EATEN REGULARLY
biscuits	S	
baked beans	S	
ice cream	S G	
chocolate	S	
jam	S G	
squash	G A	
ketchup	S	
canned cola	S A	
snack bar	G S	
tinned fruit	S	
swiss roll	S A	
corn flakes	S	
fruit pies	S A	
bread	S	
tinned soup	S	

4 Look at the table below. It shows the amount of sugar produced for the European Community by foreign countries under special agreement. Draw bar charts showing present levels of production, and colour them red.

COUNTRY	PRODUCTION (tonnes)	thousand tonnes 50 100 150 200 250 300 350 400 450 500
Mauritius	present 490,000	
	future	
Fiji	present 165,000	
	future	
Guyana	present 159,000	
	future	
Jamaica	present 118,000	
	future	
Swaziland	present 117,000	
	future	
Barbados	present 50,000	
	future	
Trinidad	present 44,000	
	future	
Belize	present 40,000	
	future	
TOTAL	present	
	future	

5 Add up the total amount of sugar produced at present.

6 Put the future total needed if there is a drop of 320,000 tonnes.

7 The cuts could be achieved if each country produced 40,000 tonnes less. Is this the fairest way? Can you think of other solutions? Decide future levels of production for each country. Draw bar charts to show these figures and colour them blue.

© Scottham, Bridge, Jewson, 1991. Macmillan. *Enquirybase Geography, Book 3*

COMMUNICATION ACTIVITY 6

MULTINATIONALS

The previous activity (Commodities) considered the exploitation of raw materials. This exercise takes the study a stage further by analysing the role of multinational companies. There has been a staggering growth of larger companies over recent years in response to consumer demand. It is estimated that three hundred giant corporations now control two-thirds of the world economy. The turnover of the largest of them exceeds the entire GNP of countries such as Denmark and Austria. Pupils are introduced to this topic through a survey of food and household goods available in local shops. The results show that despite an apparent abundance of choice, one large company, Unilever, dominates the market.

Skills
Conducting a questionnaire
Using tally marks
Using percentages
Drawing conclusions

Attitudes and values
Is it healthy for so much power to be concentrated in the hands of a few giant companies? You could discuss who might be adversely affected. Does anybody benefit? Consider especially the case of developing countries. Do multinationals bring investment and prosperity or do they simply exploit the poor?

Lesson preparation
1 Before interviewing other people, pupils should record which products they use themselves. This will make them familiar with the questions.
2 There are ten products in the survey, and pupils conduct ten interviews. Thus they should receive one hundred replies altogether, so the total for Unilever products automatically becomes a percentage.

Local enquiry/homework
1 Make your own survey of goods in a local shop or supermarket. Find out what goods are supplied by Procter and Gamble, another multinational company, by making a list of different brands of washing powder, toothpaste and soap.
2 Working from the list of Unilever products on the activity sheet, make a collection of advertisements. Is there any evidence that these products are made by the same company? Say why you think Unilever gives the impression that the goods are in competition.

Extended investigation
Find out about other multinational companies such as Shell, IBM, ICI and Exxon. Do they specialise in one product, controlling all aspects of its manufacture and distribution, or do they trade in a variety of goods? Where do the different companies have their headquarters? How do they ensure that they produce their goods as cheaply as possible? Make a sample study.

Problem
List ten ways you would try to create brand loyalty for a new product made by a multinational company.

MULTINATIONALS

A few powerful multinational companies control much of the world's trade.

1 Make a survey of ten pupils in your class. Think about each product listed below. For each one, ask them to choose the *one* brand they regularly use, and record their answer with a tally mark.

TOOTHPASTE	Crest	SR	Signal	Close Up	Colgate	Other (none)
WASHING POWDER	Persil	Daz	Surf	Tide	Omo	Other (none)
SOAP	Lux	Knight's Castile	Lifebuoy	Palmolive	Camay	Other (none)
DEODORANT	Mum	Amplex	Sure	Impulse	Right Guard	Other (none)
MARGARINE	Flora	Summer County	Stork	Echo	Blue Band	Other (none)
TEA	Brooke Bond D	PG Tips	Typhoo	Twinings	Liptons	Other (none)
FROZEN FOOD	Ross	Findus	Bejam	Bird's Eye	Iceland	Other (none)
ICE CREAM	Wall's	Lyon's Maid	Loseley	Horton's	Fiesta	Other (none)
MEAT PRODUCTS	Matteson Wall's sausages	Fray Bentos pies	John West tinned meat	Tyne tinned meat	Oxo Cubes	Other (none)
SOUP	Heinz	Campbell's	Batchelor's	Knorr	Crosse and Blackwell	Other (none)

3 Add up the totals for Unilever products.

2 All the brands listed below belong to Unilever, a multinational company. Look back at the table and colour all the Unilever products.

SR	Lifebuoy	Brooke Bond D
Signal	Sure	PG Tips
Persil	Impulse	Bird's Eye
Surf	Flora	Wall's
Omo	Summer County	Oxo
Lux	Stork	Matteson Wall's
Knight's Castile	Echo	Fray Bentos
	Blue Band	Batchelor's

4 Colour one square in the shopping trolley opposite for each Unilever product.

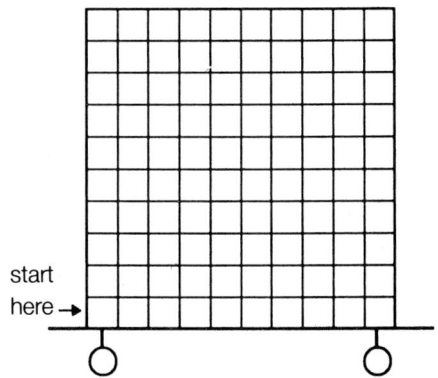

start here →

5 What percentage of household goods are supplied by Unilever?

6 What type of goods does Unilever dominate?

© Scoffham, Bridge, Jewson, 1991. Macmillan. *Enquirybase Geography, Book 3*   ACTIVITY 6

PLACES ACTIVITY 7

DIFFERENT HOUSES

Buildings are a visible record of how human beings have responded to their environment. In the earliest times, people struggled to survive against hostile forces. Then, as buildings improved, they lived for a while in harmony with nature. Now the balance has shifted still further. People have put up buildings over wide areas, creating, as in the case of London, significant microclimates. In this activity pupils compare a tropical rainforest, desert and temperate (British) climate. They then see how it has affected traditional building types. The exercise also raises questions about resources, aesthetics and cultural expectations.

Skills
Analysing climate statistics
Making comparisons
Relating climate to buildings
Making annotated sketches

Attitudes and values
Many new buildings in capital cities throughout the world are now constructed in an anonymous 'international' style. Will greater uniformity help to bring people closer together, or should we actively seek to promote diversity? You might also discuss the importance of buildings and architecture in our everyday lives. Is it worth paying extra money for fine buildings, or is the cheapest solution always the best?

Lesson preparation
1 Pupils should identify Colombo and Khartoum on the world map before beginning the activity. It is also important that they have some knowledge of the main world climatic regions.
2 The drawings in Question 2 show traditional house types, as they are much more influenced by climate than modern buildings.
3 Question 4 could be completed as a homework exercise.
4 Pupils will need statistics for world climatic regions in Question 5.

Local enquiry/homework
1 Calculate the mean temperature and rainfall for each of the places in the activity sheet.
2 Make a bar and line graph of the climate of your area. Write notes describing the most important features and try to assess how the climate has affected the local economy and settlement pattern.
3 Make a study of local architecture. Using annotated sketches, compare one building that seems to respond to climatic conditions with one that appears to ignore them.

Extended investigation
Compare a number of differing climatic regions. Has the climate influenced traditional house building? How have modern buldings overcome these forces? Is modern architecture completely successful in dealing with hostile environments? Consider special problem areas such as the tundra or earthquake zones. What influences does traditional style have on the modern architect? Make a sample study of famous new buildings in different parts of the world.

Problem
How well does your home deal with each season's weather?

Source: *Architecture for Beginners*, Louis Hellman, Writers and Readers, 1986

DIFFERENT HOUSES

The style and character of traditional buildings reflect the climate.

1 Look at the statistics for temperature and rainfall in Colombo and Khartoum. Using this information, answer the questions below and colour the graphs.

COLOMBO (Sri Lanka) Tropical Rainforest

Temp °C	25	26	26	27	26	26	27	28	28	27	27	26
Rainfall (mm)	120	310	390	150	90	110	200	310	240	110	60	95
	J	F	M	A	M	J	J	A	S	O	N	D

KHARTOUM (Sudan) Desert

Temp °C	21	23	26	29	32	31	30	29	30	28	26	23
Rainfall (mm)	0	0	0	0	5	10	50	70	15	5	0	0
	J	F	M	A	M	J	J	A	S	O	N	D

Lowest temp. _____ Highest temp. _____

Temp. range _____ Wettest month _____

Lowest temp. _____ Highest temp. _____

Temp. range _____ Wettest month _____

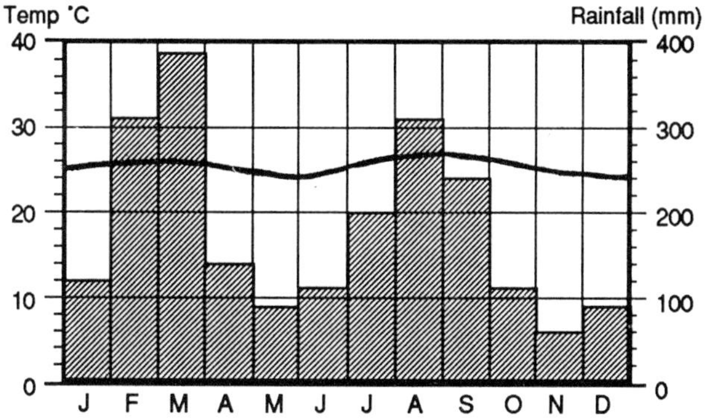

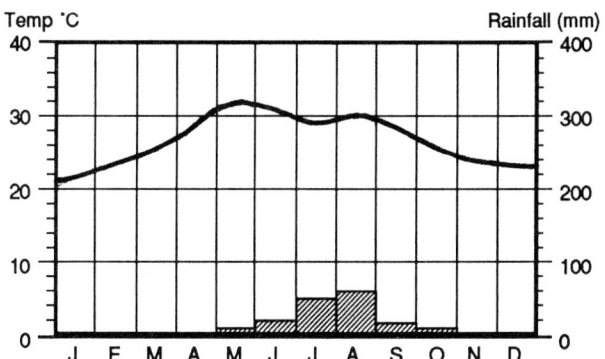

2 Say how each of the features labelled in the drawings is the result of the climate.

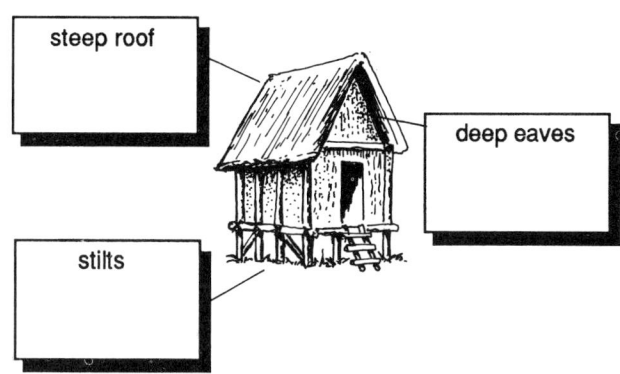

steep roof | deep eaves | stilts

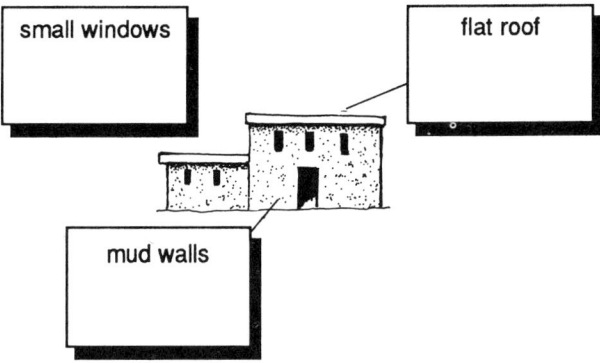

small windows | flat roof | mud walls

3 Answer the questions about the temperature and rainfall in Birmingham, and draw a graph in the same way as before.

Temp °C	5	5	7	10	13	16	17	16	13	10	8	6
Rainfall (mm)	40	35	40	40	45	50	60	60	50	65	60	55
	J	F	M	A	M	J	J	A	S	O	N	D

Lowest temperature _____

Highest temperature _____

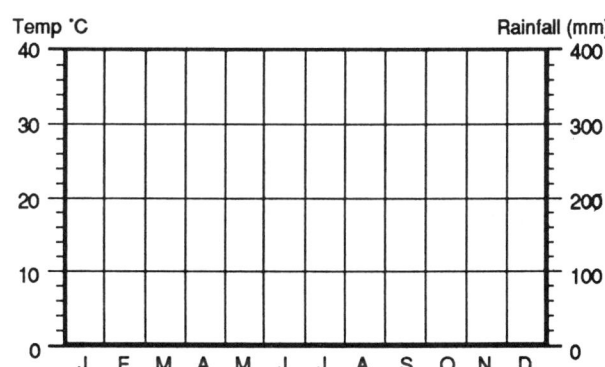

4 On a separate sheet of paper make a drawing of a house in your own area. Label the building features which have been influenced by the climate.

5 Draw some more line and bar graphs of climates in different parts of the world.

© Scoffham, Bridge, Jewson, 1991. Macmillan. *Enquirybase Geography, Book 3*

PLACE ACTIVITY 8

TOWN OR COUNTRY?

The first towns date back to neolithic times when settled forms of agriculture created enough surplus food to support an urban population. Since then towns have played a continuous and increasingly important role in human affairs. However the distinction between town and country, which was once extremely clear, has now become blurred. Better transport and communications mean that people can live in rural areas while pursuing urban lifestyles. The proliferation of small, specialist firms and the development of information processing has amplified the trend. Pupils are introduced to the idea of the urban/rural continuum in this activity. They classify the grid squares on two sample maps and then make a similar analysis of their own area.

Skills
Using a colour code
Reading a plan/map
Using a grid
Writing a description

Attitudes and values
As towns have grown larger, people have become increasingly concerned to preserve the countryside. Are current measures adequate, or do they need to be improved? Is the countryside important anyway, or are there other more pressing problems in your locality?

Lesson preparation
1 Discuss the terms used in the key. It is important that pupils understand the different categories, such as 'suburban' and 'urban/rural fringe'.
2 Pupils should colour the maps as lightly as possible so they can still be read.
3 Some discussion may be needed when deciding on the most suitable description for the grid squares on the Durham map.
4 Question 4 is designed to give pupils practice in map reading and constructing a transect.

Local enquiry/homework
1 Take a more detailed look at your locality. Working from a 1:10,000 map, classify the grid squares in the same way as before. Then calculate the number of houses in each square and try to determine the density level which is typical of each zone.
2 Compare the two sample grid squares. Make a survey of the different types of housing in each one – detached, semi-detached, terraces, flats and bungalows. What conclusions can you draw?

Extended investigation
Find out about Green Belts. When did they originate? What is their purpose, and have they been successful? What other attempts have there been to limit town growth, e.g. development 'corridors' and new towns? Make a sample study of ribbon development, using a local example if possible. What were the advantages and disadvantages of this form of building? Consider the way towns depend on the surrounding countryside. How is this evident in the large cities of the Third World?

Problem
Devise a logo that illustrates the character of your area.

Between the wars, ribbon development spread out into the country along many arterial roads.

TOWN OR COUNTRY?   Some landscapes are neither urban nor rural.

1 Using the key, select a suitable description for each grid square on the school plan.

KEY

SCHOOL PLAN	MAP OF DURHAM	CODE
completely built up	urban (town centre)	grey
mostly built up	suburban (regular rows of houses)	blue
half built up	urban/rural fringe (enlarged villages)	yellow
new development	estates (often in open country)	orange
mostly open	countryside (scattered housing)	green

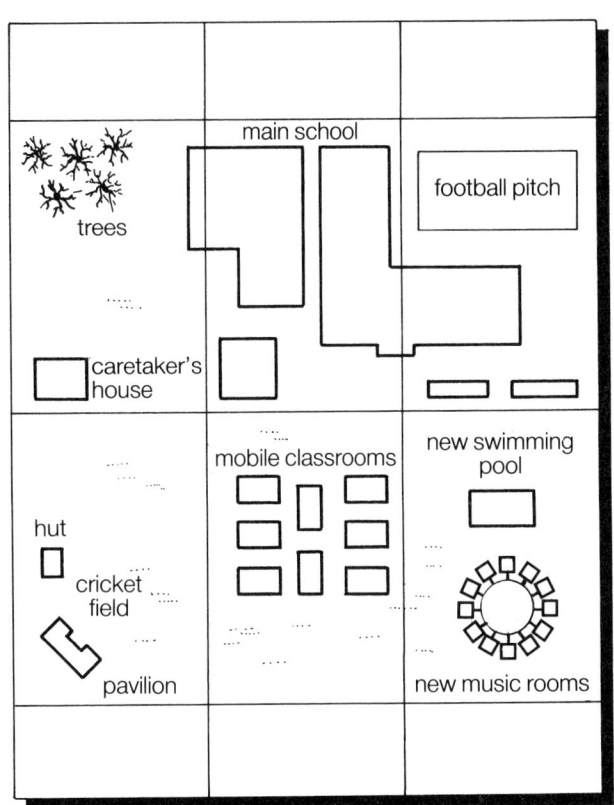

2 Colour the plan using the codes from the key.

3 Now choose a suitable description for the grid squares on the map of Durham. Colour them using the same system.

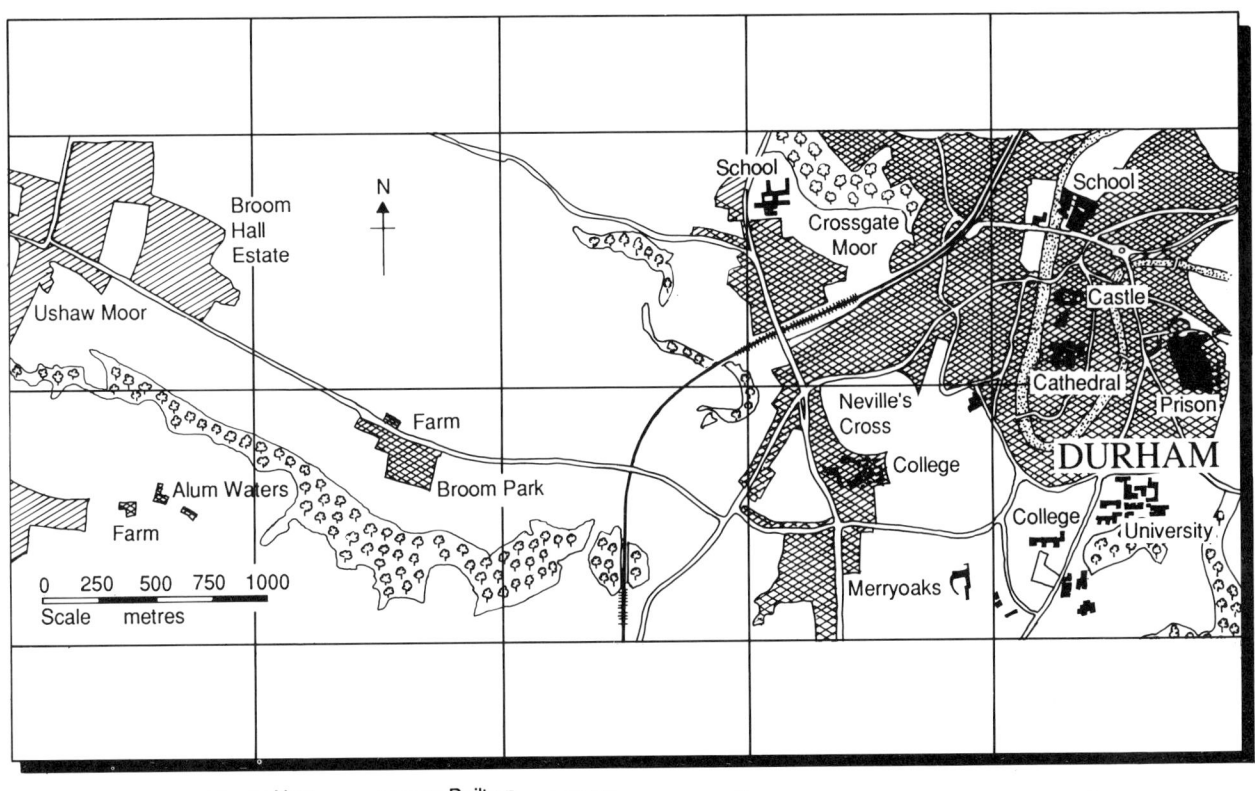

KEY: New estates, Built up areas, Woods, Roads, Railway, River

4 Imagine you are travelling from Ushaw Moor to Durham University via Broom Park and Merryoaks. Describe the different landscapes you would pass, using the descriptions from the key.

© Scoffham, Bridge, Jewson, 1991. Macmillan. *Enquirybase Geography, Book 3*   ACTIVITY 8

PLACE ACTIVITY 9

DESERTIFICATION

We live in an age in which people have suddenly discovered that they have the power to control important aspects of the environment. One of the consequences is to unleash indirect forces that are bringing far-reaching and unexpected changes. For example, fresh water pollution is increasing by 2 per cent a year. Acid rain and the greenhouse effect are beginning to have vast ecological consequences. Soil erosion and desertification are threatening large areas of the earth's surface. In this activity pupils are introduced to environmental change through a study of front gardens. Working from plans, they create a grid and classify the land use in each square. They then draw a line graph showing how the proportion of productive land is steadily decreasing.

Skills
Drawing a plan/map
Working in the immediate environment
Drawing a line graph
Making predictions

Attitudes and values
Why does desertification matter? What parts of the world are worst affected? Whom does it most concern? How could it be prevented? Discuss the problem, using sample studies by way of illustration.

Lesson preparation
1 It is helpful to discuss the table before beginning the sheet, as it shows how to distinguish between productive and non-productive land.
2 Pupils can add plans of their own gardens from memory. However, the activity can be easily altered to provide fieldwork experience by blanking out the three sample plans. Pupils will then be free to study four gardens of their own choice.
3 In Question 3 pupils must classify each grid square according to its use, and not just colour the plans.
4 There are a hundred grid squares in all, so the totals in the table will show percentages.

As more and more trees are felled for firewood, people are being forced to use animal dung and crop stalks for fuel. This deprives the soil of yet more nutrients and accelerates the process of erosion.

Local enquiry/homework
1 Make a similar study of your own garden at home. Identify the main land-use types on a plan and devise a grid which will enable you to calculate the percentage given over to each one.
2 Is there any evidence of soil erosion in your locality? Look for accumulations of soil at the bases of hedges, walls and other obstructions. Examine slopes on cultivated land for shallow gullies and changes in soil colour. Record the findings using photographs, annotated drawings and sketches.

Extended investigation
Look at a land-use map of Britain or Europe. Why is some land non-productive? Consider the forces causing desertification in different parts of the world – pressure of population, forest clearance, soil degradation and so on. What was the great 'Dust Bowl' disaster in the United States? Make a sample study of contemporary projects in the Sahel or northern China which are halting the encroachment of the desert.

Problem
Produce a leaflet explaining how householders in your area (including those living in flats) could increase the productivity of their land.

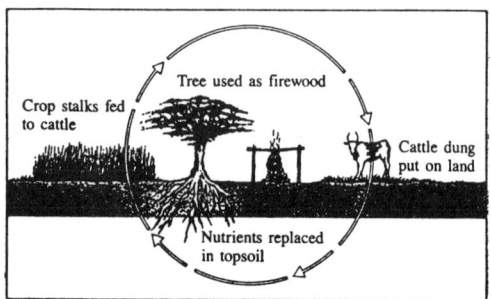
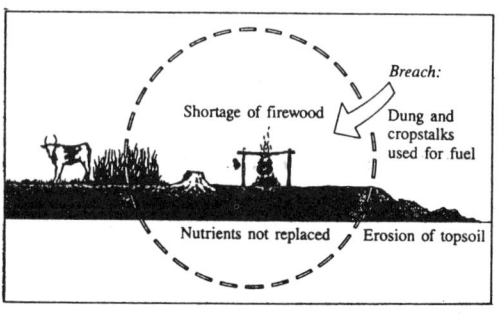

Source: *'Vanishing Earth Viewer's Guide'*, BBC, 1989

DESERTIFICATION

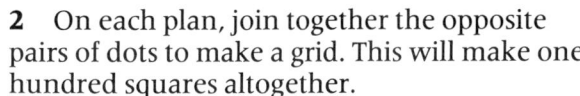

Desertification is threatening large areas of the world.

1 Look at the different garden plans below. Add a plan of your own garden in the empty space.

2 On each plan, join together the opposite pairs of dots to make a grid. This will make one hundred squares altogether.

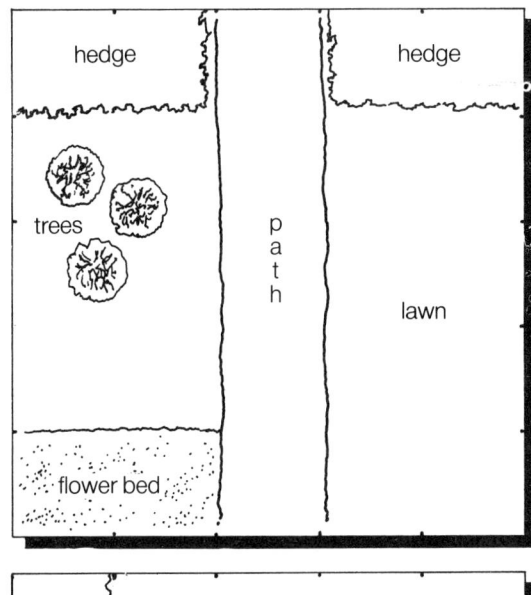

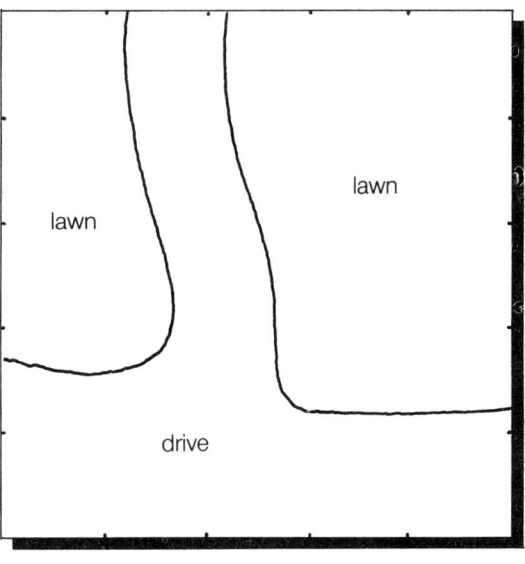

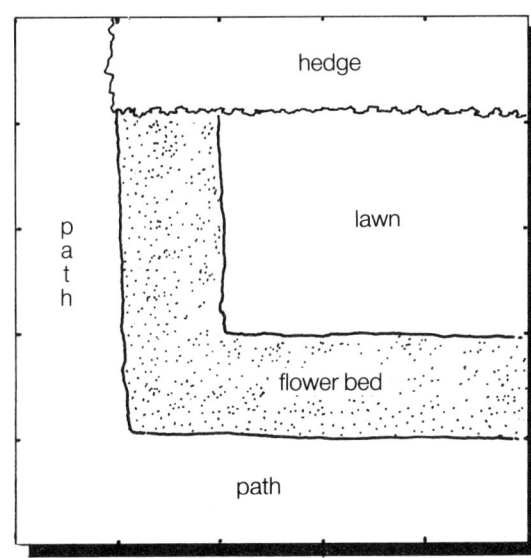

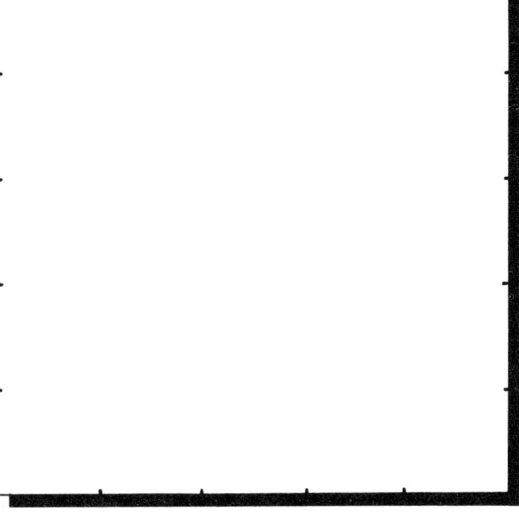

3 Decide on the main land use in each grid square. Colour them, using the code from the table.

4 Add up the totals for each land use type.

5 What percentage of the land is productive (cultivated, pasture and woodland)?

_____

6 Mark the percentage of productive land with a cross at the correct point on the graph.

7 Put a second cross showing when your area was 100 per cent productive, i.e. before it was developed.

8 Complete the graph by joining up the points, and show what might happen in future using a dotted line.

GARDEN	LAND USE	CODE	TOTAL
flower bed and vegetables	cultivated	yellow	
lawns	pasture	green	
shrubs, hedges, trees	woodland	brown	
paths, drives, ponds and walls	non-productive	grey	

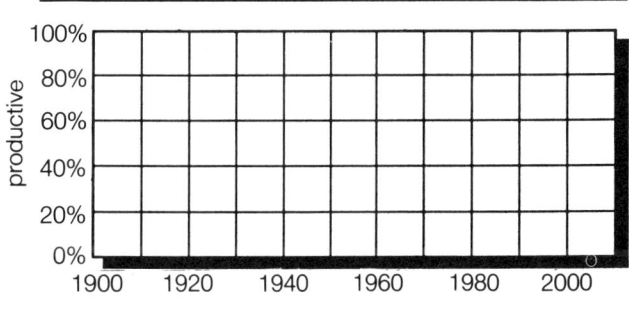

© Scoffham, Bridge, Jewson, 1991. Macmillan. *Enquirybase Geography, Book 3*

ACTIVITY 9

REGION ACTIVITY 10

PHYSICAL REGIONS

Regional geography is given less prominence nowadays than in the past, but there is still an enormous number of ways in which geographers describe different places, and in doing so establish boundaries for their descriptions. For example, urban geographers look at towns, neighbourhoods and their sphere of influence; meteorologists consider the differences and similarities that constitute a climate; geologists make detailed studies of soil and rock types in different areas. In this activity pupils define a physical region using four main criteria. Not only is the exercise presented in an exciting way using a stylised 'computer' map, it also leads to a definition of the region in which the pupil actually lives.

Skills
Reading specialised diagrams
Identifying regions
Using percentages
Writing a report

Attitudes and values
Are some physical regions 'better' than others? What type of landscape and climate do you prefer? Which parts of the world have an ideal climate? Do they have any drawbacks of a different nature? You might ask the pupils to prepare individual answers and then compare the results.

Lesson preparation
1 By way of preparation, check that the pupils understand the terms 'relief' and 'geology', and revise the climate of the British Isles.
2 The map is divided into one hundred grid squares. Each one represents one per cent.
3 Some pupils may lose their position when they transfer information to the blank map. They should number each row and work systematically down the map from north to south.
4 Pupils may need to refer to an atlas when writing their report.

Local enquiry/homework
1 Using tracing paper make a series of overlay maps of Britain showing differences in relief, rainfall, temperature and geology. On each one shade the regions which are different from your own. Then put the maps together to show the limits of your area in the same way as in the activity sheet.
2 Study a geological map of your area. What are the main rock types? Are the main landscape features the result of the underlying geology? How have they been modified by climate and other forces?

Extended investigation
Identify the major climatic regions of the world. What are their main characteristics? What is the natural vegetation in each area? Make a scrapbook collection of photographs of different landscapes. In each case describe the relief, rainfall, temperature and geology and locate the area on a map. How can human activity create new zones and regions? Consider the impact of irrigation schemes in different parts of the world. What changes might be brought about as a result of the 'greenhouse effect'?

Problem
Make a 'museum collection' of rocks, pebbles, stones and minerals from your region.

Geographers use a specialist vocabulary to describe physical regions and landscape features.

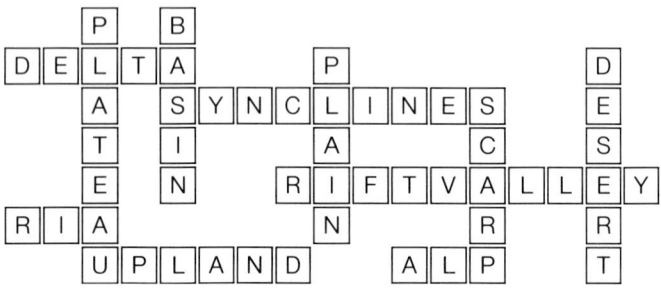

PHYSICAL REGIONS

Physical regions can be defined by relief, rainfall, temperature and geology.

1 Colour a square on each of the maps below to show where you live.

2 Circle the description under each map which best describes your locality.

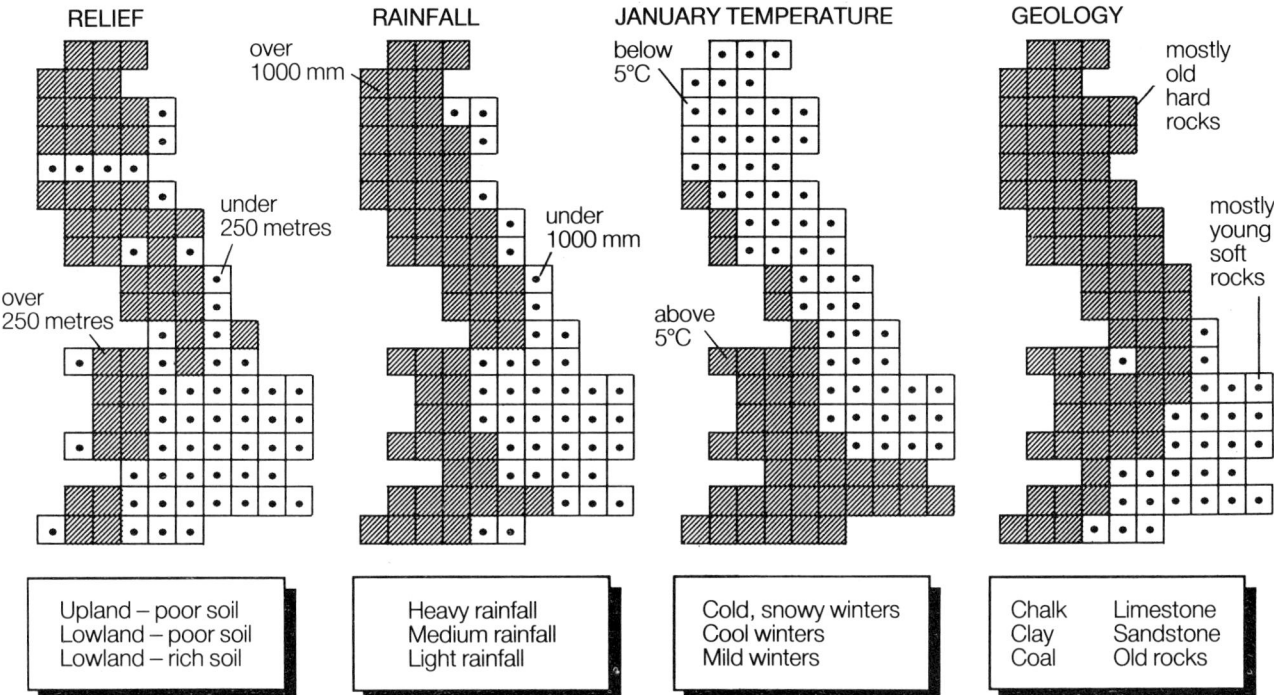

RELIEF	RAINFALL	JANUARY TEMPERATURE	GEOLOGY
Upland – poor soil Lowland – poor soil Lowland – rich soil	Heavy rainfall Medium rainfall Light rainfall	Cold, snowy winters Cool winters Mild winters	Chalk Limestone Clay Sandstone Coal Old rocks

3 Now put a cross to mark the square where you live on the map opposite.

4 Find out the size of your region. Working from the maps at the top of the page, shade all the squares which are different from yours in terms of relief. Then do the same for the areas which have different rainfall, temperature and geology.

5 Write a report describing your region. What are the chief features? What percentage of the country does it occupy? Are there any other areas of Britain which are similar?

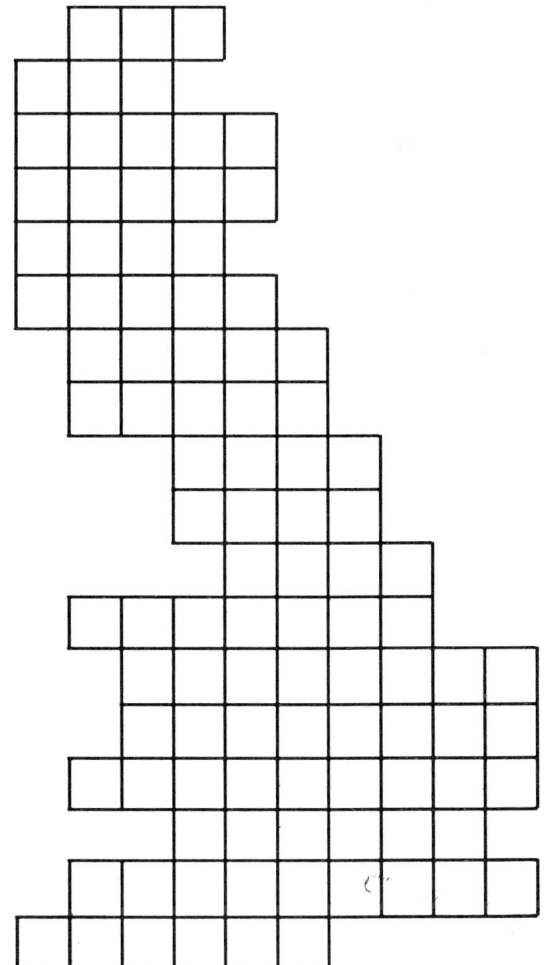

© Scoffham, Bridge, Jewson, 1991. Macmillan. *Enquirybase Geography, Book 3*   ACTIVITY 10

INDUSTRIAL REGIONS

REGION ACTIVITY 11

In the nineteenth century, Britain's wealth and power were based on a variety of heavy industries. Chief among these were coal mining, metal manufacture, textiles, chemicals, pottery and shipbuilding. The early part of the present century saw a shift towards light industry and the development of engineering and car manufacture. More recently, electronics and high-tech industries have emerged while recession and structural change have transformed the old established activities. In this exercise pupils begin by exploring the economic base of their locality through a fieldwork exercise. Working from a map, they then consider how it compares with the nearest industrial region. In the process pupils are introduced to the idea that many modern industries are still clustered around the towns and cities that were in the forefront of the Industrial Revolution.

Skills
Making a list
Identifying regions
Reading a plan/map
Making comparisons

Attitudes and values
Unemployment and economic recession have hit the industrial areas of the north and west of Britain much more severely than the south and east. Why has this happened, and how can the problem be dealt with? You might also consider the advantages and disadvantages of living in an industrial area. What are the factors which are important in determining the quality of life?

Lesson preparation
1. Information about the locality is best gathered through a fieldwork (homework) exercise. However, if necessary Question 1 could also be completed through a class discussion.
2. See that pupils are clear about the meaning of 'heavy', 'light' and 'high-tech' industries.
3. The map in the activity sheet is based on information in a school atlas. It could easily be amended to take account of any recent changes and developments.
4. Question 5 raises questions about the similarities and differences between different industrial regions. This is pursued in Question 6.

Industrial cities are not necessarily unpleasant places to live. Social factors such as crime, health and shopping facilities are key elements in creating a desirable urban environment, according to a Glasgow University research project.

Local enquiry/homework
1. Find out more about local industries. When were they established, and for what reason? Draw a map using symbols to show their location.
2. What type of industry predominates in your area? Write a report, using some (or all) of the following headings: access to raw materials, historical circumstances, local initiative, Government grants.

Extended investigation
Find out about the structure of different industries. In which ones are all the processes gathered together in a single integrated factory? In which ones is production divided between a number of different centres? Make an overlay map showing the main resources in Britain (coal, oil, iron ore and so on). How closely do they match the industrial regions? Identify the location of the main British power stations. What are the factors controlling their location? Compare especially coal, hydro-electric and nuclear generating plants.

Rankings of British cities for quality of life

1	Edinburgh	19	Bristol
2	Aberdeen	20	Derby
3	Plymouth	21	Norwich
4	Cardiff	22	Birkenhead
5	Hamilton-Motherwell	23	Blackpool
6	Bradford	24	Luton
7	Reading	25	Glasgow
8	Stoke-on-Trent	26	Bournemouth
9	Middlesbrough	27	Leeds
10	Sheffield	28	Sunderland
11	Oxford	29	Bolton
12	Leicester	30	Manchester
13	Brighton	31	Liverpool
14	Portsmouth	32	Nottingham
15	Southampton	33	Newcastle
16	Southend	34	London
17	Hull	35	Wolverhampton
18	Aldershot-Farnborough	36	Coventry
		37	Walsall
		38	Birmingham

Source: *Town and Country Planning*, October 1988

INDUSTRIAL REGIONS

Some areas of Britain are centres of industrial activity.

1 Make a list of local manufacturing industries on the table below. For each one, decide the activity, the type of industry and its size.

NAME	ACTIVITY	TYPE		SIZE	
		heavy	light/high-tech	large-scale	small-scale
		TOTAL			

2 Add up the totals in the table and circle the highest scores in the type and size columns.

3 Using this information, describe the type of industry you have found in your survey.

_____

_____

4 Now put the numbers from the table next to the main industrial regions on the map.

5 Working from the map, complete the table of industries by drawing the correct symbols in the empty boxes.

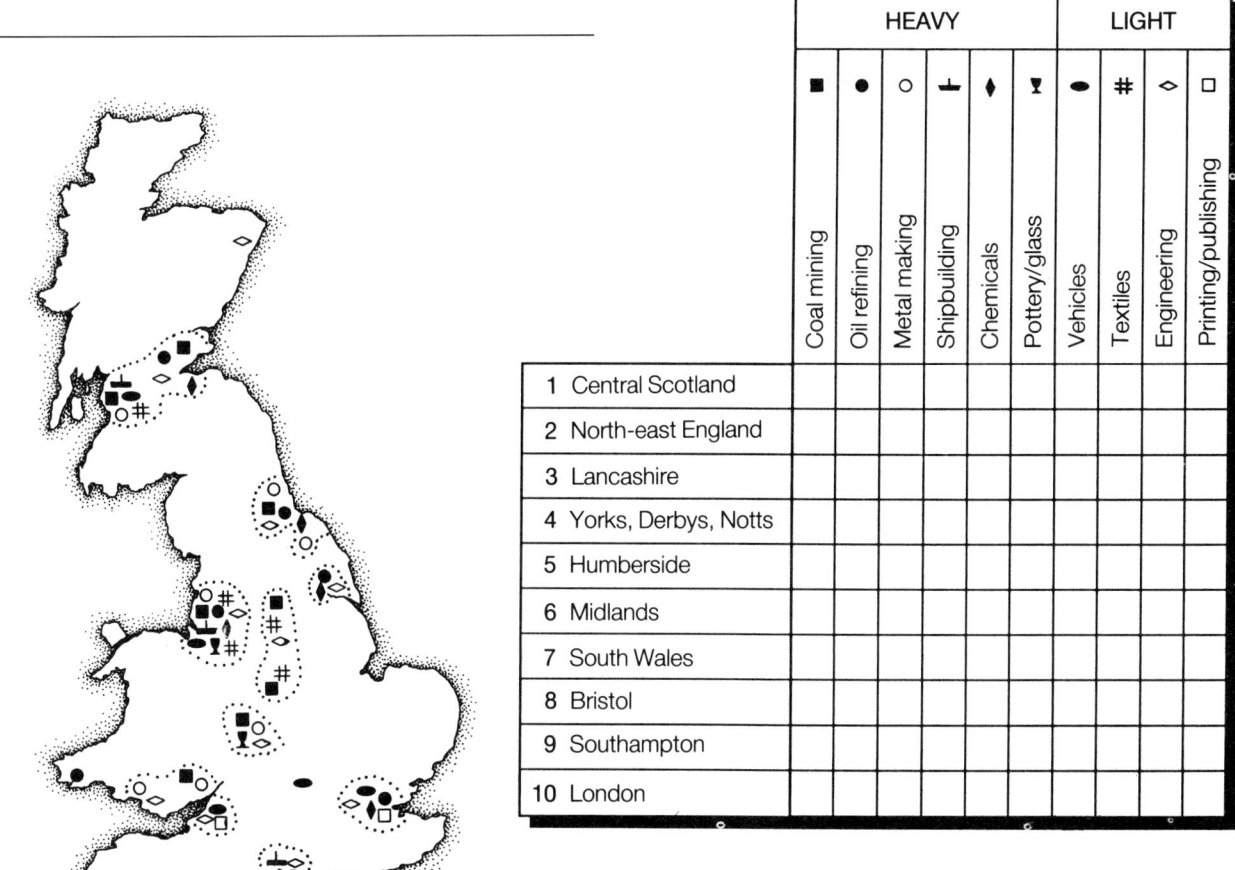

	HEAVY						LIGHT			
	■	●	○	⊥	♦	▼	⬢	#	◇	□
	Coal mining	Oil refining	Metal making	Shipbuilding	Chemicals	Pottery/glass	Vehicles	Textiles	Engineering	Printing/publishing
1 Central Scotland										
2 North-east England										
3 Lancashire										
4 Yorks, Derbys, Notts										
5 Humberside										
6 Midlands										
7 South Wales										
8 Bristol										
9 Southampton										
10 London										

6 How does the industry in your area compare with industry in your nearest major industrial region? Discuss your findings and write a report.

REGION ACTIVITY 12

SERVICE REGIONS

Modern life depends on a vast range of services and methods of distribution. These are organised over a variety of distances depending on the physical and logistical requirements of the individual system. Hospitals and local government, for instance, operate in a fairly compact area; computer and letter-based organisations over much wider areas. This activity introduces the idea of the social infrastructure. Working from a map, pupils discover the headquarters for a range of common services such as telephones, schools and water supply. The analysis investigates the optimum size for these administrative regions and proposes a model for a comparative local study.

Skills
Calculating distances using a scale
Reading specialised diagrams
Using a pie chart
Drawing conclusions

Attitudes and values
The exercise raises a variety of questions which you might discuss. For example, are large organisations more efficient than smaller ones? Should essential services be run by local or central government, or by private companies? Is it best to centralise services in a capital city, or should they be devolved?

Lesson preparation
1 Some pupils may need to orientate themselves by finding Buckinghamshire and the Thames Valley on a map of England.
2 Pupils must calculate the distance to each headquarters using the scale. The broken line indicates that Swansea and Glasgow are off the edge of the map, and the distance is therefore given.
3 In Question 3 it is valuable to discuss the distance limits for local and regional administration.
4 The pie chart should be labelled with the largest group displayed first, clockwise from twelve o'clock.

Local enquiry/homework
1 Make a list of administrative systems used in your school. Where is each one centred? Does it operate throughout the building or is it restricted to a particular place such as the library? Record your answers on a table.
2 What evidence can you find of different administrative authorities in your locality? Make sketches of any building or other clues that you discover. In what way is your area important?

Extended investigation
What administrative systems control the movement of air and sea traffic around the British Isles? Where does each one have its headquarters? Where are the main military bases? Which European cities carry out administrative functions for the European Community? Make a list of international charities such as the Red Cross, Oxfam and Amnesty International. What services do they provide, how do they operate and who provides the money?

Problem
What does your Council do, and how does it affect your life?

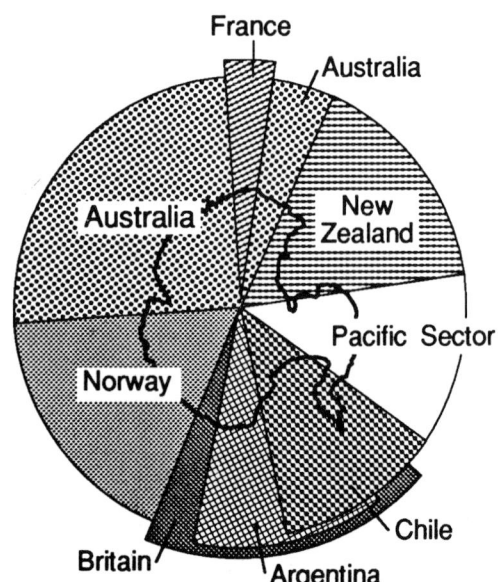

A number of countries have laid claim to Antarctica. Their zones overlap and are, at present, purely nominal.

SERVICE REGIONS

There are different administrative regions for different services.

1 John Blakemore lives in the village of Hambleden, Buckinghamshire. The table shows some of the services he uses and the authorities which provide them.

SERVICE	DISTRICT AUTHORITY	HEADQUARTERS	DISTANCE	Local	Regional	National
Schools	Buckinghamshire County Council					
Electricity	Southern Electricity Board					
Gas	North Thames Gas					
Water	Thames Water					
Income tax	Inland Revenue					
Car tax	Ministry of Transport					
Poll tax	High Wycombe District Council					
Hospital	Oxford Regional Health Authority					
Telephone	Thameswey District					
			TOTAL			

2 Using the map, find the headquarters for each service and calculate its distance from Hambleden.

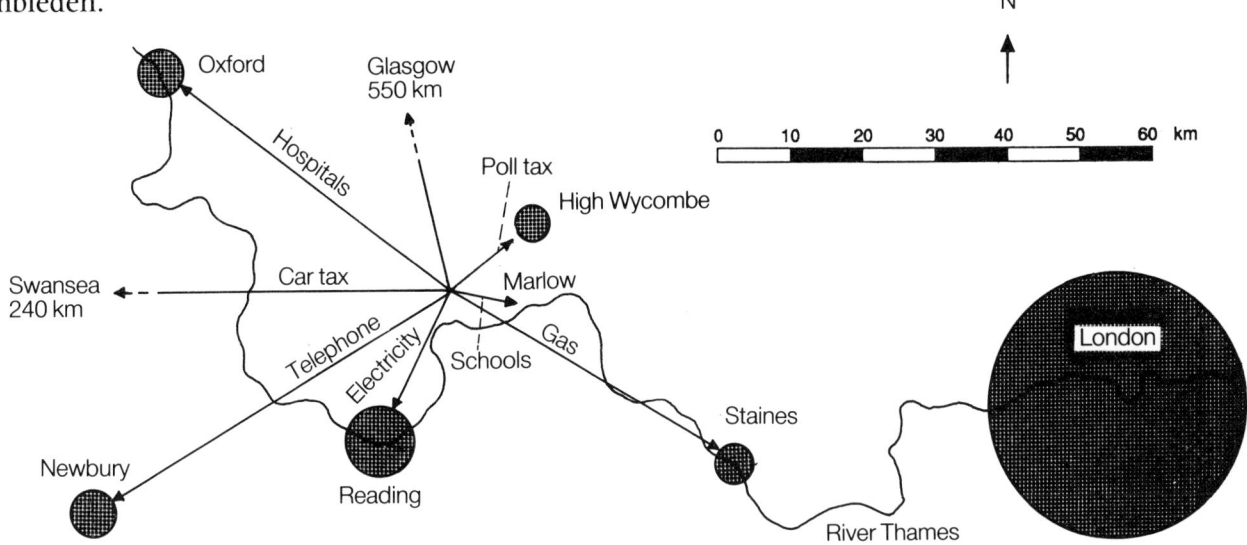

3 Decide whether each service is organised on a local, regional or national level. Tick the correct column on the table and add up the totals.

4 Show your results on the pie chart opposite. Remember to start with the biggest total, colour the sectors and label them.

5 From the survey, which type of administrative region is the most common?

6 Working from a telephone directory, make a similar survey of service regions in your locality. Draw them on a map and label the headquarters.

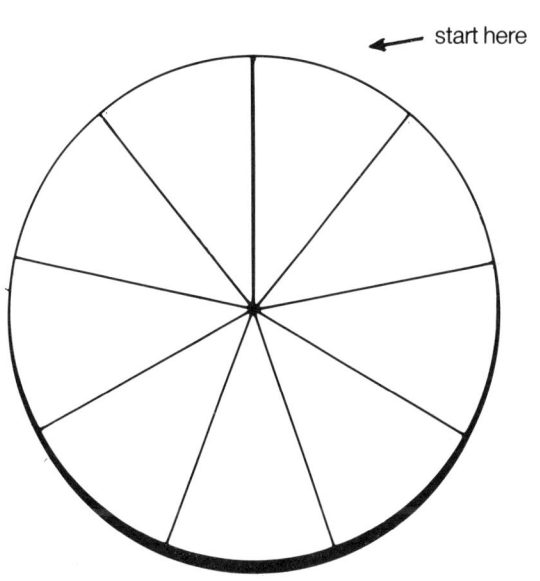

© Scoffham, Bridge, Jewson, 1991. Macmillan. *Enquirybase Geography, Book 3*   ACTIVITY 12

RESOURCES ACTIVITY 13

SPARE RESOURCES?

The resources of a region are the total means which people can use to meet their needs. Traditionally geographers have concentrated on studying physical resources such as food and raw materials. However, a broader definition would include buildings, human skills and transport and communication systems. In this activity pupils analyse how the different rooms in their school are used throughout the year. They then consider if they could be put to more effective community use – a question which is particularly relevant at a time when local financial management is giving individual schools greater power and autonomy in controlling their own affairs.

Skills
Making a list
Tabulating data
Considering alternatives
Making value judgements

Attitudes and values
Schools and other public buildings are specialist places designed for a specific purpose. Is it right to put them to greater community use? Would you need to impose any regulations? What extra costs might be involved? Consider also the question of political groups. Would some activities need to be prohibited?

Lesson preparation
1 Before beginning, discuss how the school buildings are currently used. The classroom, for example, will probably be used seven hours a day, two hundred days a year. Pupils will need to calculate the use of other rooms.
2 Question 5 will require discussion. Ask the pupils to consider who might use the school in the evenings, at weekends and during the holidays.

Local enquiry/homework
1 Make a survey of other buildings and facilities in your area which could be put to better use by the community. At what times could they be available, and how might they be used? Make notes of any problems or difficulties you can foresee.
2 Design a community centre for use in your locality. What services would it provide, whom would it cater for and why? Try to ensure that it is used fully throughout the day, and suggest how this might be organised.

Extended investigation
What other resources in Britain are under-used? Make a sample study of how the railways or canals could be put to better use. Consider human resources. Are there any projects or schemes which are held back by lack of suitable personnel? Think about new inventions and discoveries. Are these shared equally? In what ways could communication satellites benefit all nations rather than just a few?

Problem
What resources (skills, talents, products) could your school sell to the general public?

1 Sunlight
2 Wind and Waves
3 The Tides
4 Biomass
5 Running Water
6 Geothermal Heat

Renewable energy resources have enormous potential. In one year, for example, we receive fifty thousand times more energy from the sun than we obtain from burning fossil fuels.

Source: *The Case for Renewable Energy*, Michael Flood, Friends of the Earth, 1989

SPARE RESOURCES?

Some resources, like school buildings, might be used more effectively.

1 Look at the places listed below. Add any found in your school which are not mentioned.

2 For each place, colour the squares on the table which best describe how many hours it is used each day and how many days it is used each year.

PLACE	Number of hours used each day				Number of days used each year				OTHER POSSIBLE USES
	over 18	11–18	4–10	under 4	over 300	225–300	150–224	under 150	
Classrooms									
Gym									
Science labs									
Music rooms									
Main hall									
Playground									
Field									
Kitchen									
Library									
TOTAL									
	possible over-use	very good use	reasonable use	under-use	possible over-use	very good use	reasonable use	under-use	

3 Add up the totals and circle the highest score for the number of hours and days that your school is used.

4 What do the totals tell you about your school building?

5 Using the empty column in the table, write down how each place might be put to other uses.

6 What groups of people in the community would benefit from the changes you suggest?

7 Discuss the problems that might be caused by over-use. What are the implications of under-use?

© Scoffham, Bridge, Jewson, 1991. Macmillan. *Enquirybase Geography, Book 3*   ACTIVITY 13

LEISURE FACILITIES

RESOURCES ACTIVITY 14

The growth of the leisure industry has been one of the striking features of the post-war period. The reduction of the working week, coupled with the increase of personal mobility and car ownership, has created an enormous demand for recreation. One response has been to establish National Parks in areas of natural beauty. Another is the provision of planned leisure facilities. In this activity pupils make a detailed study of Rutland Water, Leicestershire. Using a table, they analyse the needs of different groups of people and identify possible conflicts. They then discover how these have been reconciled.

Skills
Labelling
Tabulating data
Considering different needs/points of view
Reading a plan/map

Attitudes and values
Leisure facilities can be contentious. They attract large numbers of people and put the surrounding area under pressure. You might discuss why facilities need to be provided. After all, people are quite capable of organising their own pastimes, or are they?

Lesson preparation
1 The matrix correlates all the activities. Thus pupils have to decide not only if fishing is compatible with water skiing, but if water skiing is compatible with fishing. It follows that the top half of the matrix should be the mirror image of the bottom.
2 Pupils may need to orientate themselves by finding Rutland Water on a map of England.
3 Rutland Water was built in the 1970s primarily as a reservoir for the East Midlands. Recreational facilities have been developed gradually and now attract half a million tourists a year. Noisy activities, like water skiing, have been deliberately excluded.

Local enquiry/homework
1 Make a study of a park or leisure area in your neighbourhood. Begin by drawing a plan. Then list the different pastimes, discover their 'compatibility index' and say what facilities have been provided for each one.
2 Using information from the public library, list the leisure facilities in your area. Which activities and sections of the community are catered for best?

Extended investigation
How have working conditions changed over the last fifty years? Consider the impact of cars and television on leisure activities. How is the environment affected? Make a study of a conservation group such as the RSPB or the National Trust. What facilities are required to host major sporting events such as the Olympic Games? Find out about traditional games and pastimes. Where did chess, nine men's morris and other games originate?

Problem
Design a pamphlet advertising the leisure facilities for young people in your area, giving as much information as you think is needed.

The use of the countryside for leisure pursuits has increased enormously in recent years.

Source: *The Effluent Society*, Methuen, 1971

LEISURE FACILITIES

Different leisure activities need to be well planned to mix together.

1 Label the leisure activities shown in the picture.

2 Put a (+) or (−) in each square in the matrix to show which activities mix together well.

3 Add up the total number of (+) to find the 'compatibility index'.

4 Which activities have the highest score? Explain why.

_____

_____

5 Which activities have the lowest score? Explain why.

_____

_____

	fishing	water skiing	rowing	bird watching	sailing	swimming	TOTAL
fishing	▨						
water skiing		▨					
rowing			▨				
bird watching				▨			
sailing					▨		
swimming						▨	

6 Look at the map of Rutland Water, Leicestershire. Colour the map and key.

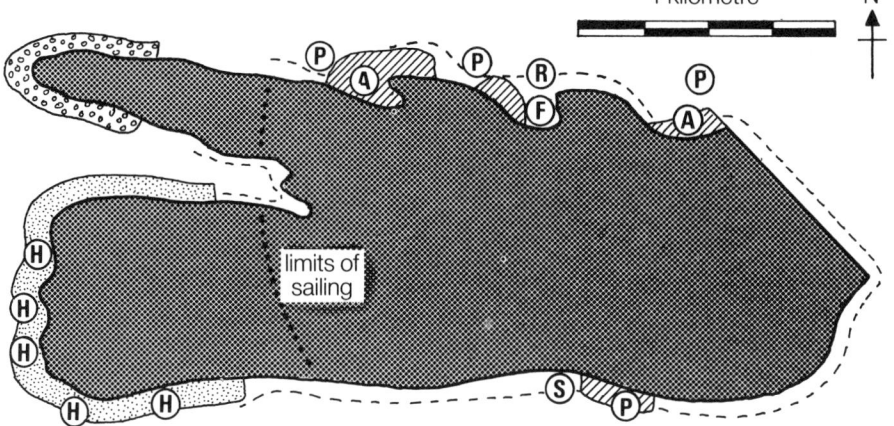

KEY

water	■
picnic areas	▨
research areas	∘∘∘
nature reserves	⋯
car parks	Ⓟ
fishing lodges	Ⓕ
adventure playground	Ⓐ
hides (bird watching)	Ⓗ
footpaths/cycleways	----
rescue centres	Ⓡ
sailing centres	Ⓢ

7 For each activity, say what facilities have been provided. Note any restrictions or controls.

ACTIVITIES	FACILITIES/CONTROLS
sailing	
bird watching	
picnicking	
fishing	

© Scoffham, Bridge, Jewson, 1991. Macmillan. *Enquirybase Geography, Book 3*   ACTIVITY 14

TOURIST FACILITIES

RESOURCES ACTIVITY 15

More and more people are going on holiday each year and many of them are going abroad. Tourism brings many benefits. It provides employment and money for regions which would otherwise be underdeveloped. This is particularly important in some parts of the Third World that are short of hard currency. However, it can also have less desirable effects such as economic dependence and cultural imperialism. In this activity pupils make a survey of holidays taken by some of their class-mates. They then deduce what facilities are required both now and in the years ahead. This leads to a fieldwork study of tourism in the immediate locality.

Skills
Making a survey
Extracting information from a table
Making comparisons
Identifying trends

Attitudes and values
Many tourists are attracted to remote and exotic locations, yet by going there destroy the very thing they set out to find. Is it true that tourism always spoils a place? Are some more vulnerable than others, and are there ways in which damage can be mitigated?

Lesson preparation
1 When making the survey pupils should enter themselves first. This will help to introduce them to the range of questions.
2 The empty spaces in the table are for recording any responses not already specified.
3 If there are two equal totals, pupils must circle them both.
4 Question 6 could be expanded into a special topic or project.

Local enquiry/homework
1 Look at old guide books and brochures for your area. Select some passages and descriptions which you think are interesting and discuss them with the rest of the class.

"Unfortunately, there was a famine for the whole two weeks we were there"

Source: *Punch*, January 1983

2 Draw a map showing the main tourist attractions within a suitable radius of your home (e.g. 25 kilometres). Add codes indicating whether each one is of historic, scenic, sporting or general interest.

Extended investigation
Find out about the history of tourism. When did towns like Bath and Brighton develop? What was the 'European tour' and to whom did it appeal? When did package holidays first become important? Make a sample study of the way tourism has affected particular parts of the world such as southern Spain or the islands of the Caribbean. What are the benefits and drawbacks for the country concerned? Trace the story of a tour operator such as Thomas Cook. How has their business changed and developed? In what ways may tourism develop in future?

Problem
Where would you locate a new tourist information centre in your area?

About one third of British holidays are taken abroad. The percentages visiting different countries in 1986 are shown in the table

	%
Belgium or Luxembourg	1
France	12
Germany (Fed. Rep.)	3
Greece	10
Irish Republic	3
Italy	5
Netherlands	2
Austria	4
Spain	35
Switzerland	2
All in Europe	88
United States	4
Other countries	6
No one country for four or more nights	2
Total	100

Source: *Social Trends* 18, HMSO, 1988

TOURIST FACILITIES

The facilities provided for tourists must match their needs.

1 Complete the survey by asking eleven people about their last main holiday.

2 Add up the totals for 'country', 'time spent', 'purpose' and 'accommodation'. In each section circle the highest score in red and the lowest score in blue.

NAME OF PERSON	PLACE(S) VISITED	COUNTRY		TIME SPENT				PURPOSE				ACCOMMODATION		
		Britain	abroad	weekend	one week	fortnight	over a fortnight	beach	touring	activity/sport	family/relations	hotel	self-catering	camping/caravan
	TOTAL													

3 Using the scores you have circled in red, describe the most common holiday and the facilities required.

FACILITIES REQUIRED	
Transport	
Accommodation	
Leisure facilities	

4 Using the scores you have circled in blue, describe the least common holiday and the facilities required for it.

FACILITIES REQUIRED	
Transport	
Accommodation	
Leisure facilities	

5 In the light of your findings, circle those items in the list below which you think are most likely to be needed by tourists in the years ahead.

railways   airports   ferry terminals   motorways

country parks   museums   disco   adventure sports

family amusements   historic buildings

shopping precincts   cinema   self-catering cottages

hotels   bed and breakfasts   camping (caravan) areas

6 Make a study of how your town or area caters for tourists. What facilities are available? Is there enough accommodation? Which times of the year are busiest? How could tourist visits be better managed?

© Scoffham, Bridge, Jewson, 1991. Macmillan. *Enquirybase Geography, Book 3*   ACTIVITY 15

PATTERN ACTIVITY 16

MICROCLIMATES

Pattern is a key geographical idea. Geologists, for example, look for patterns and configurations in the earth's surface. Urban geographers look for patterns in towns and cities. Planners make predictions based on human behaviour. Meteorologists infer weather patterns from past experience. This activity introduces the idea of small-scale weather patterns through the study of microclimates. Pupils begin by drawing symbols on a school plan and identifying different regions. They then consider how the microclimate might be altered and carry out a similar survey in their own environment.

Skills
Reading a plan/map
Using symbols
Identifying regions
Making predictions

Attitudes and values
Through their activities people create microclimates around their homes and buildings. Is it right, however, to seek to alter the climate on a larger scale? How could changes be agreed, and who would benefit? Is it always possible to predict the consequences?

Lesson preparation
1 Introduce the idea of microclimates by discussing local examples.
2 Pupils should colour code the plan lightly so as not to obscure the details.
3 Question 4 requires imaginative thinking. For example, some of the school buildings might be demolished or redeveloped, the new trees will mature and the old ones might be replaced.
4 Pupils will need an outline plan of their school or teaching block in Question 5.

Local enquiry/homework
1 Analyse your home and other local buildings to see how they create different microclimates. Use notes and drawings to record the effect of open patios, trees, ponds, conservatories, blank walls, and so forth.
2 Set up a number of 'weather stations' at selected places in and around your school. Record the temperature and daily rainfall. Show the results on a map or plan.

Extended investigation
What is the influence of microclimates in different parts of the world? Consider coastal breezes, frost pockets, south-facing slopes and other examples. Study how people have set up their own microclimates, e.g. shelter belts to protect orchards and irrigation schemes to water crops. In what other ways is human activity affecting climate? How can natural events, like volcanic explosions, bring about short-term changes to the weather? How can Landstat photographs and thermal maps help to detect microclimates?

Problem
Suggest how four microclimates in your school might be put to positive use (e.g. by building a wind generator in a draughty spot).

Diagram of the heat island over the centre of Reading on a summer morning. (It requires a wind of over twenty knots to properly ventilate a city the size of London.)

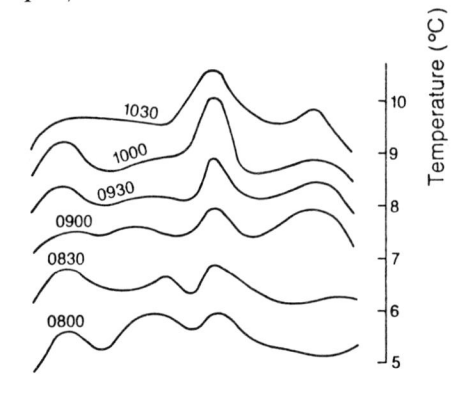

Source: *Systematic Geography*, Brian Knapp, Allen and Unwin, 1986

MICROCLIMATES

Buildings and vegetation can affect local weather conditions.

1 Look at the plan of St Stephen's Comprehensive School. It is on a flat site and the prevailing winds come from the south west.

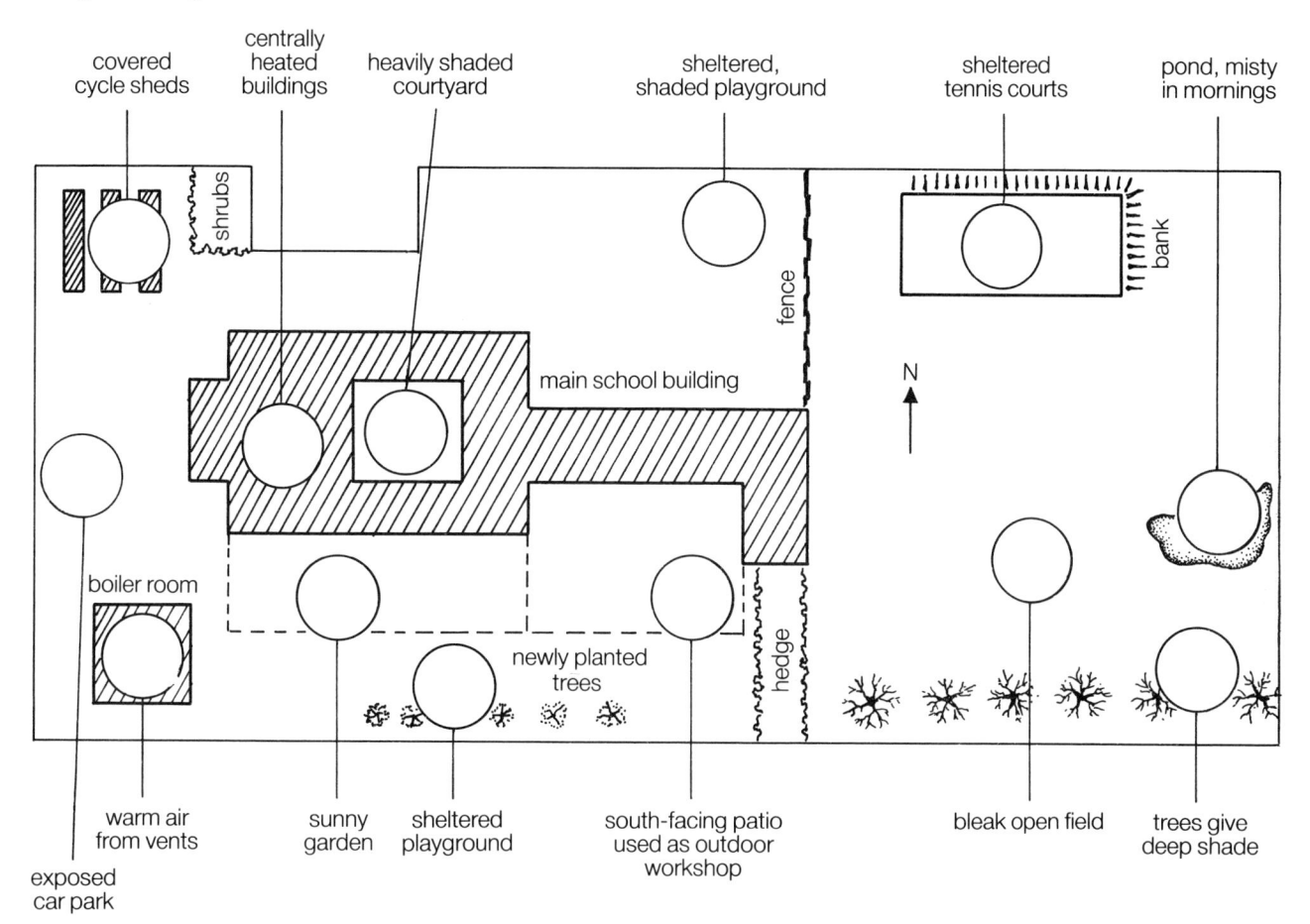

[Plan of St Stephen's Comprehensive School showing the following labelled features:]

- covered cycle sheds
- centrally heated buildings
- heavily shaded courtyard
- sheltered, shaded playground
- sheltered tennis courts
- pond, misty in mornings
- shrubs
- fence
- bank
- main school building
- N (north arrow)
- boiler room
- newly planted trees
- hedge
- warm air from vents
- sunny garden
- sheltered playground
- south-facing patio used as outdoor workshop
- bleak open field
- trees give deep shade
- exposed car park

2 Show the different microclimates by drawing symbols from the key on the correct part of the plan.

3 Starting with the school building, decide on the approximate boundaries for each microclimate. Colour each one using the codes from the key.

4 Make a list of how the school might change and say how this would affect the microclimate of the school.

MICROCLIMATE	SYMBOL	DESCRIPTION OF REGION	CODE
strong winds	↗	cold	Blue
open to forceful, driving weather	☂	exposed	Blue
local mist and fog	☁	damp	Blue
heavily shaded	///	cool	Green
dry, protected	⌂	sheltered	Green
warm and sunny	☼	hot	Yellow
artificially heated	🔥	dry	Yellow

POSSIBLE CHANGES	EFFECT ON MICROCLIMATE

5 Using an outline plan of your school, carry out a similar survey to identify microclimates.

© Scoffham, Bridge, Jewson, 1991. Macmillan. *Enquirybase Geography, Book 3*

ACTIVITY 16

PATTERN ACTIVITY 17

DISRUPTION

In the modern world people are dependent not just on each other but on a wide range of systems and services as well. Most of these are designed for a regular pattern of use. If one of them breaks down, it can have a knock-on effect and unexpected consequences. In this activity pupils work out what might happen if the caretaker fails to arrive at school one morning. They then consider some of the other events which would disrupt the school. The exercise highlights the fact that people with organisational responsibility need to be reliable, and stresses the value of predictable behaviour patterns in a complex institution such as a school.

Skills
Making a list
Identifying systems
Making predictions
Tabulating data

Attitudes and values
Human beings are not machines and they are bound to make errors and mistakes. Should people be held responsible when things go wrong, or should we offer them help and sympathy? If we pay them money to do a job, can we expect them to be more reliable and dependable? How else can they be encouraged?

Lesson preparation
1 Pupils may be puzzled that no problem occurs when the caretaker fails to take his morning break. However, this is only to be expected as he is not working at this particular moment.
2 In Question 4 pupils could include natural events like floods and snow in their list of problems.
3 The idea of systems is explored in more detail in Enquirybase Geography Book 2

Local enquiry/homework
1 Make a list of local repair and breakdown services, using the Yellow Pages and other sources. Which type of repair system seems most in evidence? Can you explain this?
2 Working from a bus or train timetable, record the number of services in your locality for each hour of the day. Explain the pattern that emerges and say at what time a breakdown would be most critical.

Extended investigation
In what way does Britain depend on other parts of the world for goods and raw materials? Consider the disruption that might be caused by the interruption of just one supply system, such as oil. What is the impact of financial factors such as inflation? Make a scrapbook of newspaper articles about different types of disruption. What seems to cause the most problems – natural events (fog and frost), accidents or deliberate human action?

Problem
Draw a network diagram of all the people and creatures that depend on you in a normal day.

Famine in Uganda 1980

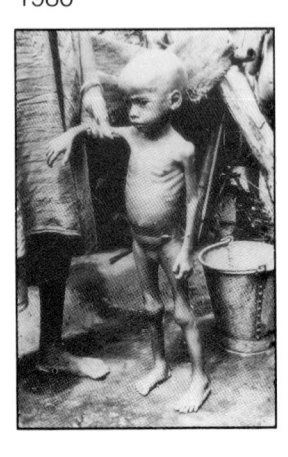

Earthquake in Peru 1971

Source: *UNICEF News*, 109, 1981

Many of the world's worst disasters have happened in modern times.

Disaster	Location and date	Deaths
Circular storm	Ganges Delta Islands, Bangladesh, November 1970	1,000,000
Flood	Huang-Ho, China October 1887	900,000
Earthquake	Shensi Province, China, January 1556	830,000
Landslide	Kansu Province, China December 1920	180,000
Volcanic eruption	Tambora Sumbawa, Indonesia, April 1815	92,000
Avalanches	Yungay, Huascarán, Peru May 1970	18,000

DISRUPTION

Many people are affected when the pattern of everyday life is disrupted.

1 Look at the caretaker's daily timetable in the table below. What might happen if he doesn't come to school in the morning? Write your answers in the 'problem' box.

2 Make a list of the people who would be affected by the disruption.

3 For each person, decide on the level of disruption using the key below.

hardly disturbed ★
disturbed ★ ★
very disturbed ★ ★ ★

CARETAKER'S TIMETABLE	PROBLEM
6.00 a.m. opens back door and lets cleaners in	
6.30 a.m. shuts off alarms and opens all doors	
7.45 a.m. teachers begin to arrive and go to rooms	
8.00 a.m. sweeps hall and puts out staff chairs	
8.45 a.m. opens playground gates	
9–10.00 a.m. breakfast break	
10–11.00 a.m. checks toilets and heating system	
12.00 a.m. sets out dining-room tables	
1–2.00 p.m. lunch break	
4.30 p.m. closes playground gates	
5–6.00 p.m. supervises school cleaners	
6.00 p.m. opens school hall for badminton classes	
9.00 p.m. locks school and sets alarms	

PEOPLE AFFECTED	DISRUPTION LEVEL
	☆ ☆ ☆
	☆ ☆ ☆
	☆ ☆ ☆
	☆ ☆ ☆
	☆ ☆ ☆
	☆ ☆ ☆
	☆ ☆ ☆
	☆ ☆ ☆
	☆ ☆ ☆
	☆ ☆ ☆
	☆ ☆ ☆
	☆ ☆ ☆
	☆ ☆ ☆

4 Look at the list of some other things which could affect your daily routine. For each one, say how your life would be disrupted and note down any other consequences. Then add some examples of your own.

PROBLEM	DISRUPTION TO YOUR TIMETABLE	OTHER CONSEQUENCES
Headteacher absent		
Form teacher absent		
School dinners not cooked		
Telephone out of order		
Fire alarm rings		
Power cut (no electricity)		

5 Consider some of the different systems in your locality. If they broke down, whom might they affect and with what consequences?

© Scoffham, Bridge, Jewson, 1991. Macmillan. *Enquirybase Geography, Book 3*

ACTIVITY 17

PATTERN ACTIVITY 18

TRENDS

One of the jobs of city planners is to make forecasts and predictions. It is their estimates which help to determine whether or not money is invested in the infrastructure such as roads, reservoirs and sewers. Schools are a significant part of this provision. The need for new buildings and resources is indicated by a detailed study of trends in the local population. In this activity pupils study the number of children in three separate schools. Working from block graphs they estimate the probable population in five years' time and consider the factors causing a school roll to rise or fall.

Skills
Identifying trends
Making a bar graph
Making predictions
Drawing conclusions

Attitudes and values
There are many issues associated with this activity. For example, should a school be allowed to take all the pupils that apply to it, even if others in the district are threatened with closure? What makes people decide to have a large or a small family? Is it right for the Government to promote or control population growth?

Lesson preparation
1 The block graphs begin with figures for the fifth year so that the trend is visible graphically.
2 You could save time and trouble by obtaining figures in advance for the number of pupils on the roll of your own school.
3 Question 5 is designed to stimulate discussion. Although St Thomas Metford has more pupils than Rosemere, it is more likely to close as its roll is falling.

Local enquiry/homework
1 What do you think are the minimum and maximum sizes at which your school could operate? Write a report assessing the sports, library and toilet facilities as well as classroom space.
2 Using Census information and local reports, find out about trends in your area. Make a detailed study of one topic such as transport, recreational facilities, housing or shops.

Extended investigation
What statistics are prepared on a daily, weekly, monthly or annual basis? How are they used? Make a study of global trends using a variety of topic headings. How can technology help in the collection and processing of data? Consider how computers have made it easier to store and retrieve information. In what ways are statistics made available for public use? Who interprets them? Look for examples of how different conclusions can be drawn from the same statistics.

Problem
What ten questions would you like to ask about your town or country twenty years from now?

Statistics

Statistically
it was a rich island.
Income per capita
one million
per annum

Naturally
it was a shock to hear
half the population
had been carried off
by starvation.
Statistically
it was a rich island

A UN Delegation
(hurriedly despatched)
dicovered however
a smallish island
with a total population
of – 2.
Both inhabitants
regrettably
not each a millionaire
as we'd presumed.
But one island owner
income per annum
two million.
The other
his cook/chauffeur

shoeshine boy/butler
gardener/retainer
handyman/labourer
field nigger etc. etc.
The very same
recently remaindered
by malnutrition.

Statistically
it was a rich island.
Income per capita
per annum
one million.

C. Rajendra

Source: *Images of the World*, Dave Hicks, Centre for Multicultural Education, 1980

TRENDS
Statistics can help people make plans for the future.

1 Colour the graphs showing the number of pupils at three different schools.

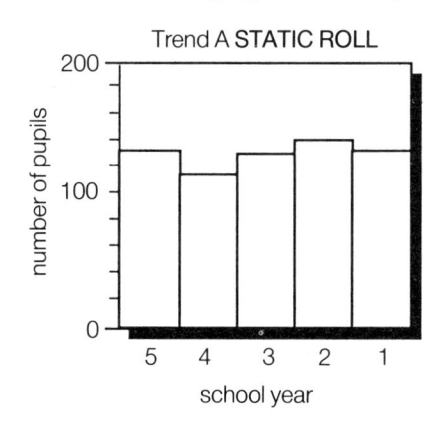

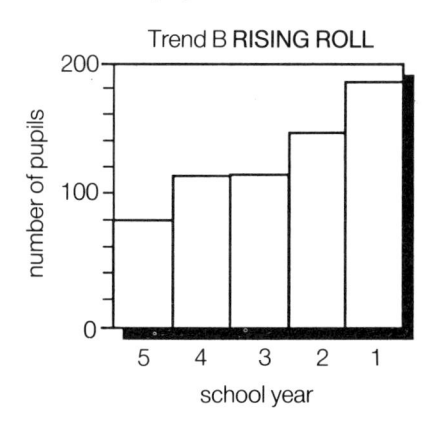

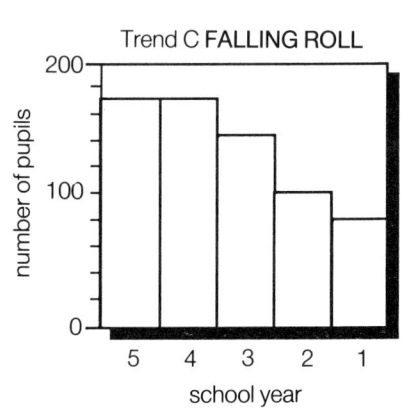

Trend A **STATIC ROLL**   Trend B **RISING ROLL**   Trend C **FALLING ROLL**

2 Now look at the statistics for the number of pupils at three other schools.
For each one, draw a bar graph and describe the trend in the space underneath.

SCHOOL	YEAR				
	5	4	3	2	1
St Thomas Metford	179	147	150	120	83
Rosemere	120	139	130	118	142
Fort Arthur	231	246	270	278	293

SCHOOL	TOTAL NOW	TOTAL IN 5 YEARS
St Thomas Metford		
Rosemere		
Fort Arthur		

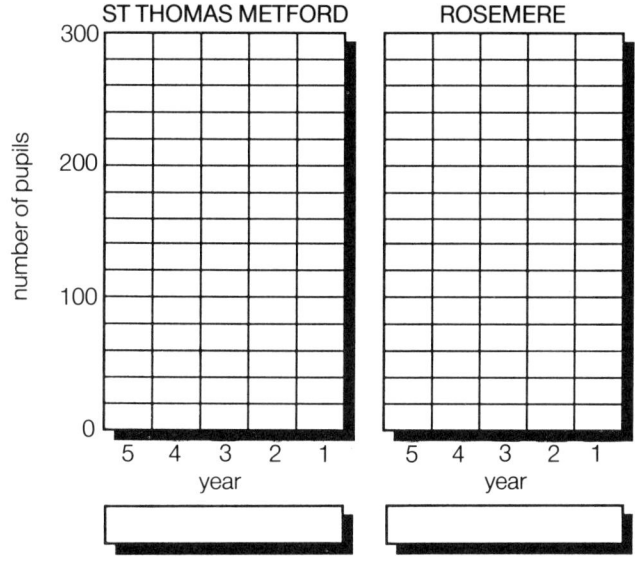

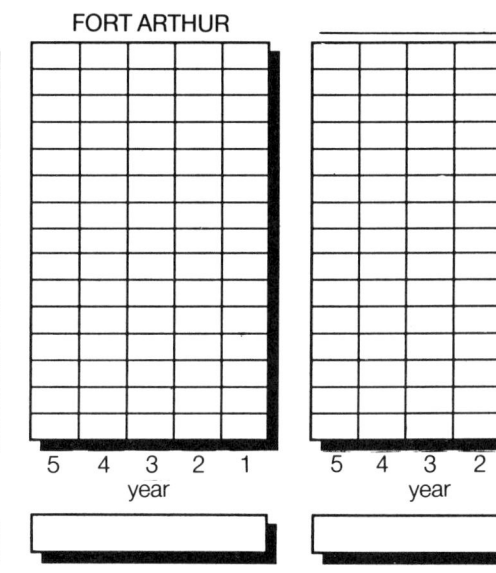

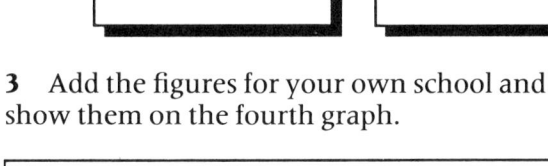

ST THOMAS METFORD   ROSEMERE   FORT ARTHUR

3 Add the figures for your own school and show them on the fourth graph.

FACTORS CAUSING FALLING ROLLS

4 Add up the numbers of pupils in each school and write the answer in the empty box above the graphs.

5 Estimate the numbers in five years' time, assuming present intake levels (i.e. the size of the first year stays constant). Which of the schools might face closure?

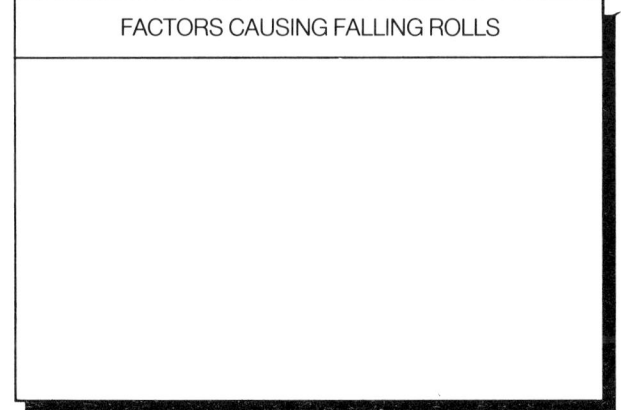

6 Make a list of the things which could cause a school roll to rise or fall.

7 Tick the factors which seem appropriate at your school.

© Scoffham, Bridge, Jewson, 1991. Macmillan. *Enquirybase Geography, Book 3*   ACTIVITY 18

DESIGN ACTIVITY 19

HOUSES GAME

Although British population growth has declined enormously since the turn of the century, there continues to be a demand for new housing. In part this stems from the declining quality of older properties. However, it is also fuelled by ever lower levels of housing occupation as more and more people decide to live on their own or in small family units. As a result there is steady pressure for development on greenfield sites, particularly in places in southern England with good access to the motorway system. In this activity pupils develop a site of their own – a blank sheet of A4 paper – using nine template houses. It is based on an original idea by Jeff Bishop and Graham Russell. Among other things it illustrates how the final design solution is usually a trade off or compromise between a number of different forces.

Skills
Following instructions
Making models
Testing ideas
Considering alternatives

Attitudes and values
New housing sites are extremely valuable. Is it right for people to profit from selling their land, and does this serve the best interests of the community? You might also consider how new development threatens rural areas. Should more money and effort be put into redeveloping towns and cities?

Lesson preparation
1 Introduce the activity by visiting new housing developments in the locality. Talk to the developers if possible.
2 Make sure the pupils have understood the rules of the game before they begin.
3 You will need to provide scissors for cutting out the templates, and blank sheets of A4 paper for the development site. Glue and modelling equipment is also useful for completing the layouts.
4 You could introduce new constraints such as safety, noise, views, wheelchairs, and energy conservation, as the game progresses.
5 Allow plenty of time for discussing the layouts. You might ask pupils to present their schemes to the rest of the class.

Local enquiry/homework
1 Find a site in your locality which could be developed for housing. Working from an accurate plan and using realistic dimensions, show how it might be developed.
2 Compare housing densities in your nearest town or settlement. Divide a 1:1,250 scale map into one-hectare squares (100 metres by 100 metres). Calculate the housing density for each square, assuming an average of four people per house. Using a colour code, show areas of high, medium and low density.

Extended investigation
Find out about current housing schemes in this country and abroad. Make a study of the London docklands redevelopment. In what way does Milton Keynes differ from other new towns? Make a comparison between low- and high-rise development. What are the problems with tower blocks? Find out about the history of building regulations. Why were they necessary? Do they still serve a useful function?

Problem
Make a model or drawing of your ideal house.

HOUSES GAME

There are many issues to consider in any new development.

1. Cut out the house shapes and assemble them.

2. Arrange the houses on the development site (a blank sheet of A4 paper). You must make the layout as convenient and attractive as possible within the following rules.

(a) There must be a footpath to every house – no-one must be cut off.
(b) Every house must have some sunshine. Mark the north point on your plan.
(c) Every house must have a reasonable amount of privacy.
(d) There must be a space, half the size of the garden, where each household can park their car.
(e) There must be a play area and as much open space as possible.
(f) You are free to draw roads, plant trees and enlarge gardens as you wish.

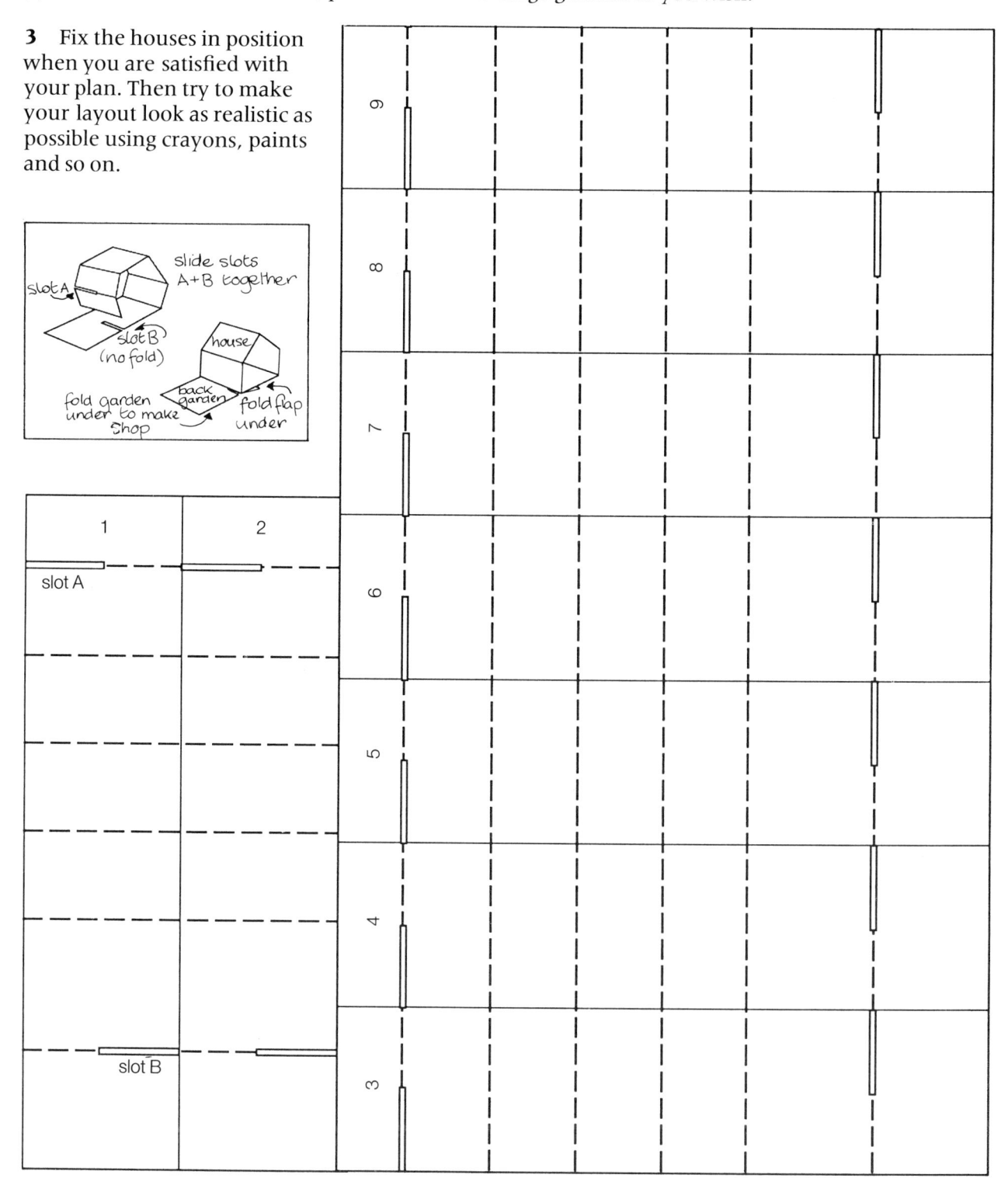

3. Fix the houses in position when you are satisfied with your plan. Then try to make your layout look as realistic as possible using crayons, paints and so on.

© Scoffham, Bridge, Jewson, 1991. Macmillan. *Enquirybase Geography, Book 3*

ACTIVITY 19

DESIGN ACTIVITY 20

PLANNING ISSUES

Before the Industrial Revolution the Government was concerned mainly with taxation, defence and trade. However, in the last hundred years it has become increasingly involved in planning issues as massive population increases and improvements in the transport system have led to more and more land-use conflicts. This activity is based on a real proposal and considers plans to reorganise a factory within its own site. Pupils are asked to study the proposals and argue in favour of the scheme against objections from local residents. This highlights how planning decisions have to strike a balance between conservation, economic development, the provision of employment and the needs of local residents who are likely to be affected by noise, traffic and pollution.

Skills
Associating plans with pictures
Using a colour code
Consider different needs/points of view
Making out a case

Attitudes and values
Discuss the reasons for current planning controls. Why should people be prevented from doing what they want on their own land? Is the present planning system stifling enterprise? The enquiry into the Sizewell B nuclear power station provides a particularly interesting case study.

Lesson preparation
1 Check that the pupils understand the four land-use categories used in the key.
2 Pupils will find the two views of the factory helpful when they colour code the plans.
3 Note that under the proposals the listed house will be demolished. Explain the listing system and how it is used.
4 The activity is based on a real planning proposal which provoked a great deal of local opposition.

Local enquiry/homework
1 Find out from newspapers, the Local Plan or Council officers what new developments are planned for your area. Show them on an annotated map.
2 Make a study of a local building or factory which has been extended or altered. Are the changes sympathetic to the site and surroundings? Are there any obvious problems with the design?

Extended investigation
Find out about major development schemes in different parts of the world. In each case, consider who benefits from the project and who loses. Is it likely to cause any damage to the environment? What makes the site suitable? Where is the money coming from? Make a study of places where development is prohibited, such as Game Reserves and National Parks. In what way is Antarctica currently protected? What might happen there in the future?

Problem
Devise a simple board game about the problems facing Gray's Paints.

Science fiction planning blight
People of the Earth, your attention please . . . as you will no doubt be aware, the plans for development of the outlying regions of the Galaxy require the building of a hyperspatial express route through your star system, and regrettably your planet is one of those scheduled for demolition. The process will take slightly less than two of your Earth minutes . . . There's no point in acting all surprised about it. All the planning charts and demolition orders have been on display in your local planning department in Alpha Centauri for fifty of your Earth years, so you've had plenty of time to lodge any formal complaint and it's far too late to start making a fuss about it now . . . What do you mean you've never been to Alpha Centauri?

Source: *The Hitch-hiker's Guide to the Galaxy*, Douglas Adams, Pan, 1979

PLANNING ISSUES

Important building alterations have to be approved by the local Council.

1 Look at the views and plans below. They show how Gray's Paints in Rochester, Kent would like to modernise their factory.

PRESENT VIEW

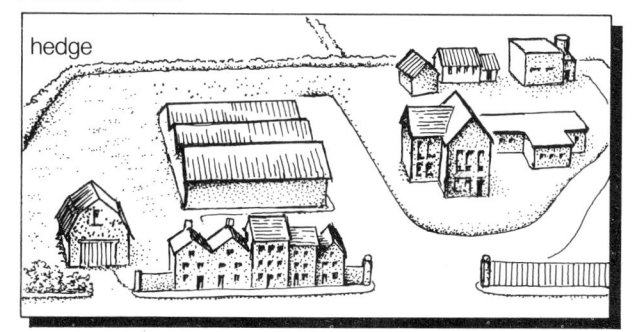

PROPOSED VIEW

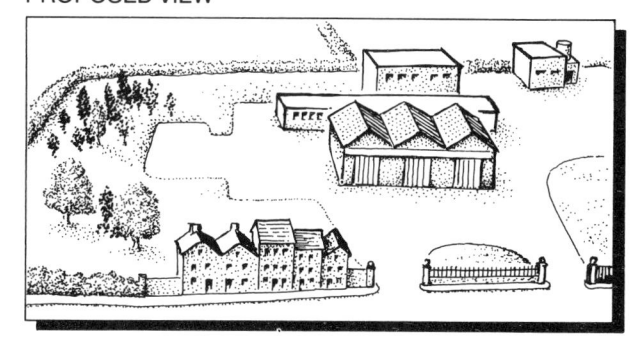

PRESENT PLAN

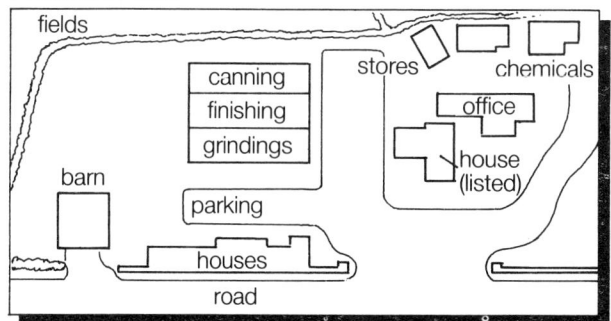

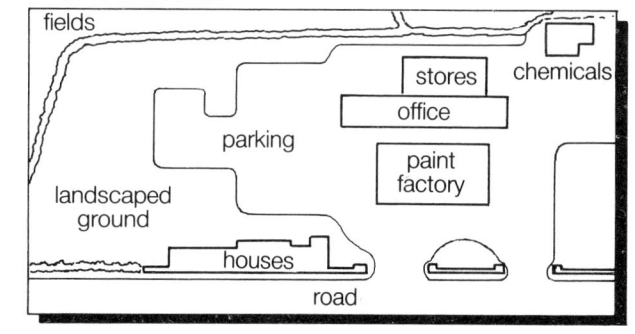

PROPOSED PLAN

2 Colour the two plans using the key below.

Residential (houses/farm buildings)	brown
Transport (roads/car parks)	yellow
Industrial (factories/offices)	red
Open land (grass/farmland)	green

3 Make a list of all the changes that you notice.

4 Some local residents object to the scheme. Here are their objections.

(a) Open land will be spoilt.
(b) There will be too many extra lorries.
(c) The new factory will be bigger and block views of the country.
(d) The landscaped ground will be used for an extension in years to come.
(e) There will be more noise from the busier factory.
(f) The 'listed' house will be demolished.
(g) The old barn with bats and other wildlife will be lost.

5 What reasons do you think the company director will put forward to convince the Council to accept the scheme?

6 Draw a better plan which would be acceptable to the company and protect the listed house, the barn and the sports field. When you have finished, list the advantages of your scheme.

© Scoffham, Bridge, Jewson, 1991. Macmillan. *Enquirybase Geography, Book 3* ACTIVITY 20

DESIGN ACTIVITY 21

A NEW BYPASS?

Many British towns are based on traditional industries. As these have declined, their economic base has shifted and become diversified. One of the visible results of this process is increasing traffic congestion – a kind of hardening of the arteries resulting from age. Rochester, Kent provides an interesting example. It was once famous for its dockyards for which the river provided a transport lifeline. Now the river exerts a stranglehold on movement around the town, funnelling traffic across a congested bridge. A new by-pass could well be the key to sustained economic regeneration, but it would be a dramatic and costly investment. In this activity pupils explore the issues involved, using a simulation.

Skills
Reading a plan/map
Assessing an improvement scheme
Considering different needs/points of view
Making out a case

Attitudes and values
Traffic congestion appears to be an insoluble problem. A new by-pass may bring temporary relief but the volume of traffic soon builds up again. In addition, any new roads are often built at the expense of the countryside and tend to encourage urban sprawl. Ask the pupils what solutions they favour and get them to explain why.

Lesson preparation
1 Pupils must study the map carefully. At present there are only two crossing points over the River Medway. Rochester Bridge in the town centre is severely congested and already carries 50,000 vehicles a day. Estimates suggest that traffic will increase by 40 per cent over the next fifteen years. Hence the need for the by-pass.
2 It may help to discuss the advantages and disadvantages of the different crossings before tackling Questions 3 and 4.
3 Questions 5 and 6 are designed to lead into a simulation.

Local enquiry/homework
1 Make a survey of traffic management projects in your area. These could vary from small-scale improvements (new kerbstones) to the construction of a major new road or by-pass. Say why you think each project was undertaken.
2 Identify the main points of traffic congestion in your locality and mark them on a map. Consider in detail how one of them might be solved, using a mixture of drawings, plans, diagrams and questionnaires.

Extended investigation
Investigate how traffic problems have been tackled in other cities such as Rome, Tokyo and Los Angeles. Compare different management schemes – special licences, tolls, controls on lorry traffic, closed city centres and so on. Make a case study of traffic congestion in London and the measures taken to alleviate it – underpasses, inner-city motorways, the M25 and other orbital systems. What alternatives to road transport may be developed in the short and long term?

Problem
When do you contribute to traffic congestion?

As well as adding to the appearance of a street, trees can help to reduce traffic noise and absorb smoke and fumes.

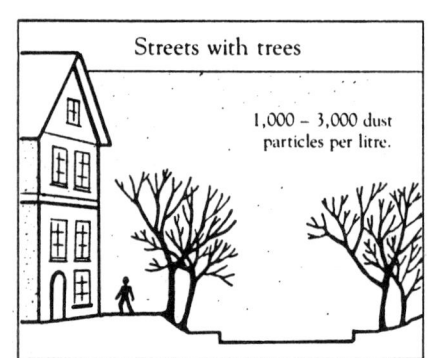

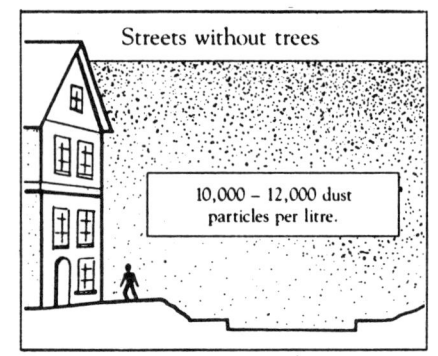

Streets with trees: 1,000 – 3,000 dust particles per litre.

Streets without trees: 10,000 – 12,000 dust particles per litre.

Source: *Task Force Trees*, Countryside Commission

A NEW BYPASS?

There are many ways of dealing with traffic problems.

1 Look at the map below of Rochester in Kent. The dotted line shows the route of a new by-pass designed to relieve congestion in the town centre and provide a new river crossing.

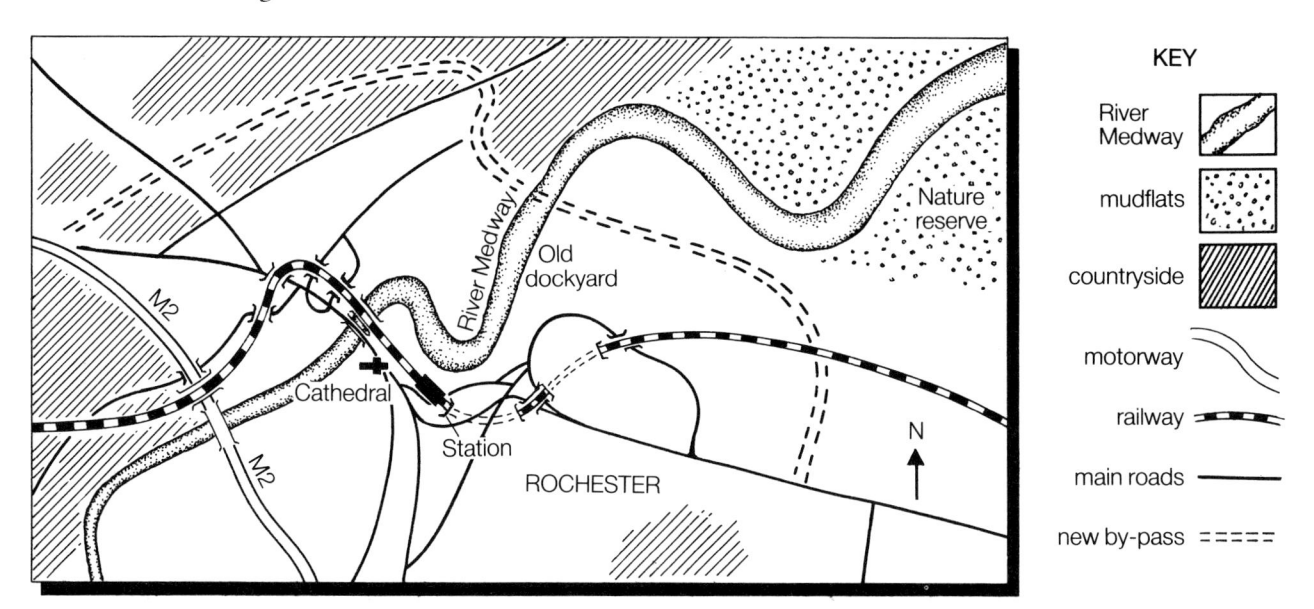

KEY
- River Medway
- mudflats
- countryside
- motorway
- railway
- main roads
- new by-pass =====

2 Complete the map and key using blue for the River Medway, yellow for the mudflats, green for the countryside and red for the M2 motorway. Mark the route of the new by-pass in brown.

3 Write down the advantages and disadvantages of the different types of crossing shown in the drawings below.

SOLUTION	COST	ADVANTAGES	DISADVANTAGES
High-level bridge	Building cost £30 million Actual running cost £100,000		
Low-level bridge	Building cost £27 million Actual running cost £200,000		
Tunnel	Building cost £42 million Actual running cost £350,000		
Do nothing Present Rochester Bridge at capacity carrying 50,000 vehicles a day	Nil		

4 Which type of crossing do you think is best? Why?

5 Select members of your class to debate the proposal from the following points of view.

a local businessman   a cyclist   a County Planner
a local resident without a car   an environmentalist
a school pupil   a delivery van driver
an old-age pensioner

6 Write a short speech for the part you are taking.

© Scoffham, Bridge, Jewson, 1991. Macmillan. *Enquirybase Geography, Book 3*   ACTIVITY 21

CHANGE ACTIVITY 22

JOB OPPORTUNITIES

Human beings thrive on a mixture of stability and novelty. They need stability for psychological security, yet because human intelligence is essentially innovative, novelty is a necessary component of a fulfilling environment. In previous centuries both these basic qualities were balanced over relatively long time scales. Now, the pace of life has accelerated. In this activity pupils look at one of the main forces behind social change – the interaction between technology and employment. Working from a list, they tabulate the jobs available to school leavers in 1919 and compare them with present-day opportunities. As well as illustrating changes in employment, the exercise raises questions about working conditions, qualifications and sex roles.

Skills
Identifying job types
Making a list
Tabulating data
Writing a report

Attitudes and values
Do you think jobs are more interesting now than they were in the past? Have working conditions improved? Do women have better opportunities? Do people receive adequate training? Discuss some of these questions with the class, using examples if possible.

Lesson preparation
1 Some of the jobs done in 1919 seem strange to us nowadays because they have disappeared. Check that pupils understand the more unusual ones such as 'wheelwright apprentice', 'pupil teacher', 'dockyard apprentice' and 'seamstress'.
2 Discussion can help when completing the list of modern jobs in Question 3.
3 Question 5 is designed to be as general as possible. It could lead to a case study of an individual family or trade.

Women have a much narrower range of jobs than men. Furthermore, in every part of the world there are low-status jobs that are specifically defined as 'women's work'.

Local enquiry/homework
1 Make a survey of the machines and equipment used nowadays in your school or home that would not have been available in 1919. Say what benefits (if any) each one has brought.
2 Make a list of jobs that are currently available in your area. How many of them may have disappeared in thirty years' time? What new jobs do you think might have taken their place?

Extended investigation
How have inventions like the steam engine and the computer brought about changes in employment? Have these always been beneficial? What are the advantages of simple, intermediate and advanced technology? Make a study of early machines by visiting an appropriate museum or heritage centre. Consider different craft industries. Why have some of them survived to the modern day? Are there any areas of Britain or Europe where high-tech industries are clustered together? Can you explain this?

Problem
Compile a 'survival pack' of skills everyone should have on leaving school.

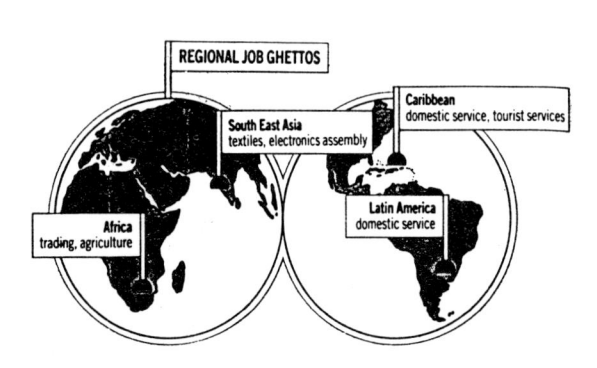

UNIVERSAL JOB GHETTOS
nursing, primary school teaching, service and sales, child care work

REGIONAL JOB GHETTOS
South East Asia: textiles, electronics assembly
Caribbean: domestic service, tourist services
Africa: trading, agriculture
Latin America: domestic service

Source: *Women in the World*, Seager and Olson, Pan, 1986

JOB OPPORTUNITIES

New technology has brought great changes to working conditions.

1 The table shows the jobs which a class of pupils entered on leaving school in 1919. Add up the dots showing the main requirements.

JOBS IN 1919

REQUIREMENTS	Butcher's delivery boy	Maid	Wheelwright apprentice	Railway porter	Apprentice carpenter	Pupil teacher	Dockyard apprentice	Army recruit	Miner	Seamstress	Farm labourer	Nurse	Apprentice hat maker	Laundry worker	Solicitor	Shop assistant	TOTAL
Male only	●		●	●			●	●	●		●						
Female only		●								●		●		●			
Open to both sexes						●							●		●	●	
Needs entry qualifications						●	●								●		
No need for entry qualifications	●	●	●	●			●	●	●	●	●	●	●	●		●	

JOBS NOWADAYS

REQUIREMENTS																	TOTAL
Male only																	
Female only																	
Open to both sexes																	
Needs entry qualifications																	
No need for entry qualifications																	

2 In the second part of the table, write down any of the jobs available in 1919 which pupils from your class might do on leaving school nowadays. Mark them with a star.

3 Fill in the gaps by writing down some modern jobs which have replaced the others.

4 Add dots showing the different requirements for each one, and add up the totals.

5 Write a short report about the changes that have occurred. How many of the 1919 jobs have been lost? In what way have job requirements changed? Is there a difference in the type of work available nowadays? Do men and women have more equal opportunities?

© Scoffham, Bridge, Jewson, 1991. Macmillan. *Enquirybase Geography, Book 3*   ACTIVITY 22

CHANGE ACTIVITY 23

ENERGY DEMAND

One of the reasons for the quickening pace of social change is the greater range of energy resources now available. Electricity in particular has loosened the ties of industry with specific places, and allowed greater mobility and flexiblity of production. This activity illustrates the extent to which we depend on electricity in our daily lives. Pupils compare their electricity needs over a twenty-four hour period with those of a sample child. They then plot the results on a line graph to show peaks and troughs of demand.

Skills
Using a scoring key
Constructing a line graph
Constructing a timetable
Drawing conclusions

Attitudes and values
Figures for electricity consumption are used to indicate the economic development of a country. What do they actually measure? Can you think of any other indicators which would be more accurate? Is it really possible to measure the quality of life anyway?

Lesson preparation
1. In Question 1 pupils will need to look at the timetable carefully in order to infer what equipment is used at different times of the day. For example, the freezer will be using electricity all the time. The fire may be switched on at intervals.
2. Remind the pupils how to draw a line graph. They must show the totals with a cross on each vertical axis, not in the gaps. They should also join up the points using a system of crosses, dots or dashes which distinguishes between Ratner Khan and themselves.

Local enquiry/homework
1. Make a similar study of the way you use water in your home. First, list what you might be doing at hourly intervals. Then record the different ways you might be using water and code them from one to three depending on the quantity required. Finally add up the totals and construct a line graph.
2. Find out how much electricity you use in your home. Take meter readings at hourly intervals throughout an evening. Draw a histogram to show your results and explain any variations that you observe.

Extended investigation
Identify industries which are prone to considerable variation of demand. Is this the result of seasonal factors, changes in fashion, daily behaviour patterns or other forces? Sort them into groups. Which industries work round the clock with teams of shift workers? Why is this necessary? Consider patterns and variations in the natural world. Make a sample study of tide levels at a given coastal location. How does the volume of water in a river vary during a year? What problems does this cause? Are there any benefits? Make a sample study.

Problem
List the different ways electricity is used in your school. Devise a question set to determine if each use is necessary or if an alternative could be found.

Average wattage of electrical equipment used in the home

Cassette deck	30	Lawn mower	250
Clock	*	Microwave oven	600
Coffee grinder	100	Radio	*
Computer	30	Record player	75
Cooker (oven & hob)	12000	Sewing machine	75
Drill	500	Shaver	16
Fan	3000	Storage heater	3000
Food freezer	300	Tape recorder	75
Grill	2500	Toaster	1360
Immersion heater	3000	Typewriter	35
Iron	1250	Washing machine	2500
Kettle	3000	Yogurt maker	25

*Consumes negligible quantities of electricity

Source: *Green Consumer Guide*, Elkington and Hailes, Gollancz, 1988

ENERGY DEMAND

There is a changing pattern of electricity demand throughout the day.

1 The timetable below shows how Ratner Khan spends her day. For each activity decide what equipment she might be using, circle the scores in the correct columns and add up the totals.

TIME	ACTIVITY	lights	TV	kettle	electric fire	cooker	radio	slide projector	freezer	TOTAL
1	sleeping	1	1	3	3	3	1	2	2	
2	sleeping	1	1	3	3	3	1	2	2	
3	sleeping	1	1	3	3	3	1	2	2	
4	sleeping	1	1	3	3	3	1	2	2	
5	sleeping	1	1	3	3	3	1	2	2	
6	sleeping	1	1	3	3	3	1	2	2	
7	waking up	1	1	3	3	3	1	2	2	
8	breakfast	1	1	3	3	3	1	2	2	
9	assembly	1	1	3	3	3	1	2	2	
10	maths lesson	1	1	3	3	3	1	2	2	
11	break	1	1	3	3	3	1	2	2	
12	drama lesson	1	1	3	3	3	1	2	2	
13	cooked lunch	1	1	3	3	3	1	2	2	
14	science lesson	1	1	3	3	3	1	2	2	
15	slide show	1	1	3	3	3	1	2	2	
16	sports practice	1	1	3	3	3	1	2	2	
17	homework	1	1	3	3	3	1	2	2	
18	dinner	1	1	3	3	3	1	2	2	
19	watching TV	1	1	3	3	3	1	2	2	
20	disco	1	1	3	3	3	1	2	2	
21	cup of tea	1	1	3	3	3	1	2	2	
22	sleeping	1	1	3	3	3	1	2	2	
23	sleeping	1	1	3	3	3	1	2	2	
24	sleeping	1	1	3	3	3	1	2	2	

2 Show these figures on the line graph below and label it.

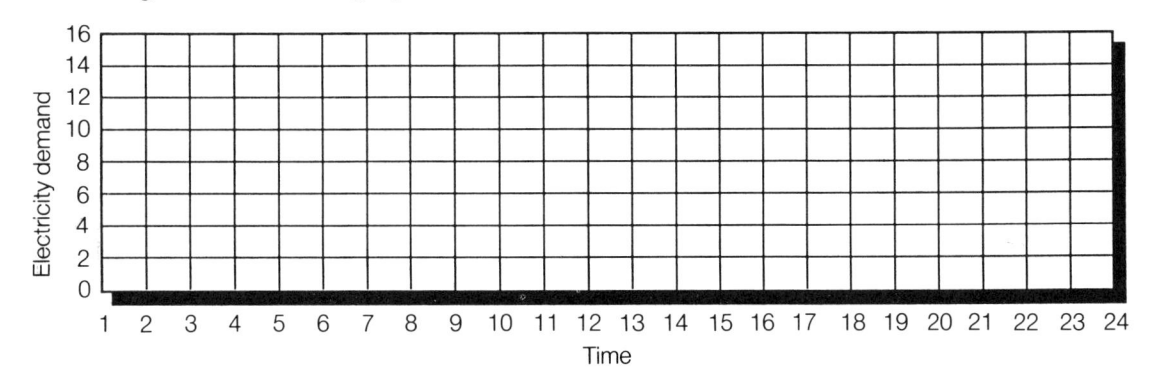

3 Now list your activities for a normal Sunday. Write down the different ways you might use electricity at each hour of the day. Give each piece of equipment a rating from 1 to 3, circle the scores and add up.

4 Draw a line graph showing your figures and label it.

5 At what times are there peaks of electricity demand?

© Scoffham, Bridge, Jewson, 1991. Macmillan. *Enquirybase Geography, Book 3* ACTIVITY 23

FORECASTING

CHANGE ACTIVITY 24

In the modern world, forecasting changes has become an important activity. Population growth and industrialisation mean that we are making ever larger demands on the planet. Non-renewable resources, for example, are being consumed at an alarming rate, pollution is degrading the global environment, new threats are appearing to human health, and ecological richness and balance are threatened. These factors cannot be dealt with on an ad hoc basis. They must be modified in the light of long-term predictions. In this activity pupils look at events in their own lives which seem predictable and consider if they are likely to be disrupted. They return to the survey several days later to assess the accuracy of their forecast.

Skills
Making a list
Considering alternatives
Using percentages
Keeping records over time

Attitudes and values
Our environment affects our behaviour and personality. How do you think continual uncertainty would affect you? Do you like unexpected events, or do you prefer to know exactly what is happening each day? Is it better to have clear-cut rules and codes of conduct, or should people make up their own minds about what is right and wrong?

Lesson preparation
1 In Question 1 pupils should list specific things they expect to do (e.g. catch 7.53 a.m. bus) rather than general events (e.g. go to school).
2 Pupils should check the accuracy of their forecast in their next geography lesson or as a homework exercise.

Local enquiry/homework
1 How does your school use forecasts? What statistics are collected? Which outside organisations are interested in the information? Write a short report.
2 Assess the accuracy of a sample weather forecast. Begin by taking a careful note of the evening forecast. Then record the weather that actually happens the following day. Finally say whether the forecast was accurate or misleading.

Extended investigation
Which international organisations use forecasts, and for what purpose? Find out about how they are used by the United Nations and major charities. How are forecasts used to predict the results of elections? Are they always right? Make a sample study of earthquakes. Where are they most likely to occur? Can forecasts help to predict the dangers?

Problem
List six ways life in school could be made more predictable.

It is difficult to distinguish between short-term weather patterns and long-term trends. However, many scientists believe that the world climate is getting warmer.

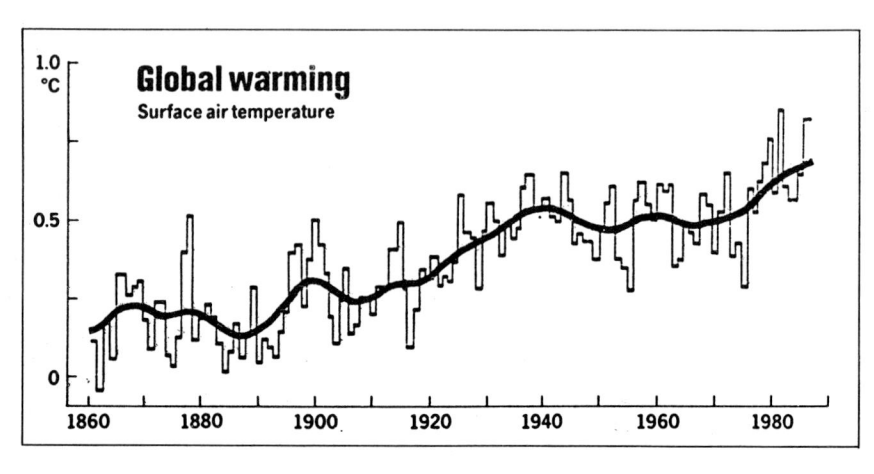

Global warming
Surface air temperature

Source: *New Statesman and Society*, July 1988

FORECASTING

Forecasters make predictions, but unknown factors may affect what happens.

1 Make a list of twenty things you expect to do tomorrow. Next to each one, say what might happen to stop you doing it.

THINGS I EXPECT TO DO TOMORROW	POSSIBLE PROBLEMS	CODE	FORECAST unlikely	possible	probable	CHECK
		TOTALS				

Multiply by 5 to discover the percentage accuracy of your forecast. → %

2 Choose a colour for each of the code boxes in the table below.

3 Using these colours, complete the code boxes in the table.

TYPE OF PROBLEM	EXPLANATION	CODE	TALLY	TOTAL
Natural events	Snow, fog, heat-wave			
Administrative changes	Other people are responsible for the change			
Personal factors	You are responsible for the change			
Health reasons	Illness			
Financial problems	You run out of money			
Communication difficulties	Messages get muddled			

4 Make a tally count of the number of times each code is used and add up the totals.

Which type of problem could cause the most difficulties?

5 Think about each of the things in your list. Complete the forecast by deciding if problems are unlikely, possible or probable. Tick the correct column and add up the totals.

6 Using your table, write a forecast describing what might happen to you tomorrow. What changes seem most probable? What other changes seem possible?

7 Discover if your forecast was accurate by returning to it in a few days' time. Put a tick in the check column for any predictions which were accurate and a cross for those which were inaccurate. Add up the total number of ticks to discover the accuracy of your forecast.

© Scoffham, Bridge, Jewson, 1991. Macmillan. *Enquirybase Geography, Book 3*

STRUCTURE ACTIVITY 25

MONEY TO SPEND

The flow of money is one of the basic dynamics of human society. In the past, towns and cities have waxed and waned according to their access to trade, commodities and raw materials. Now economics is a sophisticated and complex system underpinning the lives of people across the globe. In this activity pupils contrast the goods that they can afford with those available to youngsters of a similar age in Turkey and Egypt. This introduces the idea that money is distributed unequally between nations, that luxury goods have a greater wealth loading in some countries than in others, and that poverty is relative to the overall wealth of a particular community.

Skills
Reading for information
Tabulating data
Transferring information to a diagram
Making comparisons

Attitudes and values
The activity draws attention to inequalities of wealth in a very personal way. Is it fair that some children will never be able to afford the things that others take for granted? Do we place greater value on the things we have earned than the things we are given? Are there any reasons why children should not be sent out to work in their spare time?

Lesson preparation
1. When showing how they spend their pocket money, pupils should enter actual amounts in each column and not just put a tick or a cross.
2. Check that the pupils understand how to draw proportional circles. These increase in size along a logarithmic scale. It is the area and not the diameter which is significant.
3. In Question 5 pupils will quickly realise it is impossible for Ali Misir to save money for a Walkman radio and that it is therefore impossible to complete the table. This should be the subject of discussion.

Local enquiry/homework
1. Make a survey of high-street shops in your area. How many of them provide ordinary household goods, and how many of them provide luxury goods and services? Calculate percentages.
2. Make a survey of a dozen television or magazine advertisements. For each one name the goods that are advertised, decide whether they are essential or luxury items, and say to whom the advertisement is designed to appeal.

Extended investigation
Make a comparison between centrally planned and free-market economies. What is planned obsolescence? Is it possible for the economies of industrialised nations to continue to grow indefinitely, or will they eventually outstrip their ecological base? Collect statistics for the GNP in different countries of the world. Why are some countries so much richer than others?

Problem
How would you organise your life if people stopped using money?

	hrs/mins	
1 large loaf (white sliced)		7
1lb of rump steak		47
500g of butter (home produced)		17
1 pint of fresh milk		4
1 dozen eggs (medium size)		14
100g of coffee (instant)		24
125g of tea (medium priced)		9
1 pint of beer		12
1 bottle of whisky	2	11
20 cigarettes		22
Weekly gas bill	1	20
Weekly electricity bill	1	06
1 gallon of petrol (4 star)		34
1cwt of coal	1	32
Weekly telephone bill	1	32
Motor car licence	27	47
Colour television licence	16	07
Cinema admission		31
Long-playing record (full price)	1	37

Average time for a married man to earn a selection of goods and services in Britain in 1986

Source: *Social Trends* 18, HMSO, 1988

MONEY TO SPEND

In some privileged countries people have plenty of money to spend on luxuries.

1 Write your name and country on the table below. Then add your normal weekly pocket money and show how you spend it by answering the questions.

NAME	COUNTRY	POCKET MONEY	AMOUNT SPENT ON ESSENTIAL ITEMS					AMOUNT SPENT ON LUXURY GOODS						
			food for family meals	essential clothes	school bus fares	household bills (heating etc.)	TOTAL for essential items	entertainment	hobbies	sport	records/music	snacks (hamburgers etc.)	leisure clothes	other	TOTAL for luxury goods

2 By reading the descriptions, enter similar information on the table about how Zeytin Karakol and Ali Misir spend their money.

Zeytin Karakol

'I live in Istanbul, Turkey. In the morning before school I help serve in my father's corner shop. After school I sometimes help deliver orders to customers. I get 2000 lira (£4) for this work. I give 1500 lira (£3) to my family to help pay household bills. I spend the rest on going to the cinema.'

500 lira = £1 sterling

Ali Misir

'I live in Cairo, Egypt. My father works as a doorkeeper at an office in the centre of the city. We live in a crowded apartment block in the suburbs. In my spare time I sell nuts in the street near our home. In a good week I earn four Egyptian pounds (£1) which I give to my mother to help buy food for the family.'

4 Egyptian pounds = £1 sterling

3 Draw circles on the diagram below to show how much pocket money you, Zeytin and Ali spend each week on essential items. Fill the gap between the lines and write the person's name next to each circle.

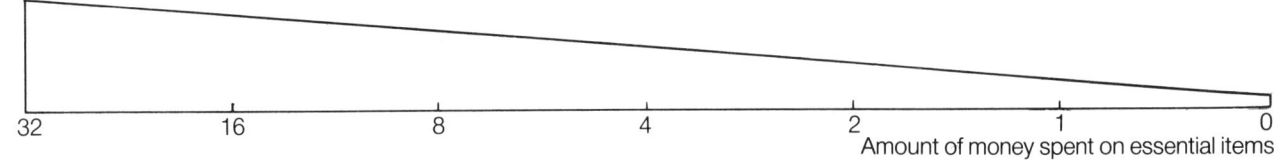

```
32    16    8    4    2    1    0
```
Amount of money spent on essential items

4 Show how much you, Zeytin and Ali spend on luxury goods in the same way and label each circle.

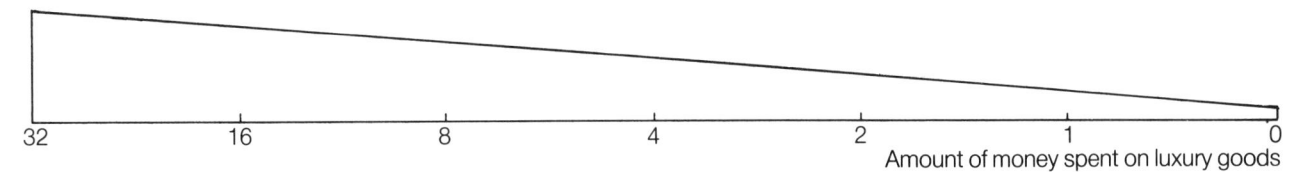

```
32    16    8    4    2    1    0
```
Amount of money spent on luxury goods

5 Colour a square in the table for each week you, Zeytin and Ali would have to save in order to buy a Walkman radio costing £45. Write the answer in the box under the table.

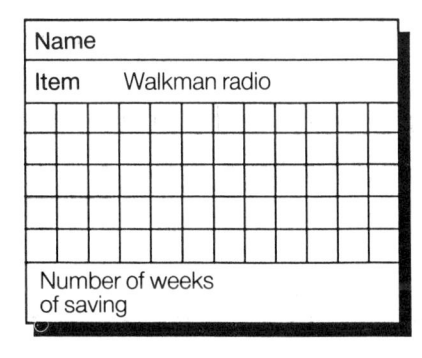

Name	
Item	Walkman radio

Number of weeks of saving

Name	Zeytin Karakol
Item	Walkman radio

Number of weeks of saving

Name	Ali Misir
Item	Walkman radio

Number of weeks of saving

© Scoffham, Bridge, Jewson, 1991. Macmillan. *Enquirybase Geography, Book 3*

STRUCTURE ACTIVITY 26

UNEQUAL SHARES

Some countries have more natural wealth than others. Britain, for example, is fortunate in having a wide range of mineral and fossil fuels, a bountiful supply of wood and water and a variety of clays and building stones. Since the seventeenth century it has also supplemented its natural wealth with resources from overseas colonies. Despite all these advantages, many people in Britain remained extremely poor until relatively recently – a situation which still prevails in many parts of the world. In this activity pupils compare their housing conditions with those of a child of similar age in Soweto, South Africa. The exercise raises questions about global inequality and the way that it can be solved.

Skills
Writing a description
Drawing inferences
Making comparisons
Using a value scale

Attitudes and values
Why do so many people live in poverty? Is it necessary? Is it right? What can we do about it? You might also consider how differences in wealth are maintained. For example, is there any evidence that great inequalities lead to violence and repression?

Lesson preparation
1 By way of preparation, it would be useful to look at photographs of housing conditions in different parts of the world.
2 Pupils might need to draft a description of their own living conditions before perfecting a final paragraph for the activity sheet.
3 Question 4 uses a value scale. The terms 'deprived' and 'privileged' have been used deliberately to provoke discussion.

Local enquiry/homework
1 Make a study of inequalities between local schools. How do the resources vary between (a) secondary schools, (b) primary schools, (c) other educational establishments?
2 Make a list of different groups of people in the local community who are deprived in some way. For each one write a sentence or two describing the problems they face.

Extended investigation
Find out about living conditions in shanty towns in different parts of the world. What are the main problems facing people who live in them? Is there any prospect that they will be solved? Make a study of slums in Britain in the nineteenth century. How did they evolve? Consider regional differences in Britain today. Why are some areas prospering while others languish?

Problem
Do you share everything in your house equally? Write a teenage survival guide.

Countries of the world according to their GNP

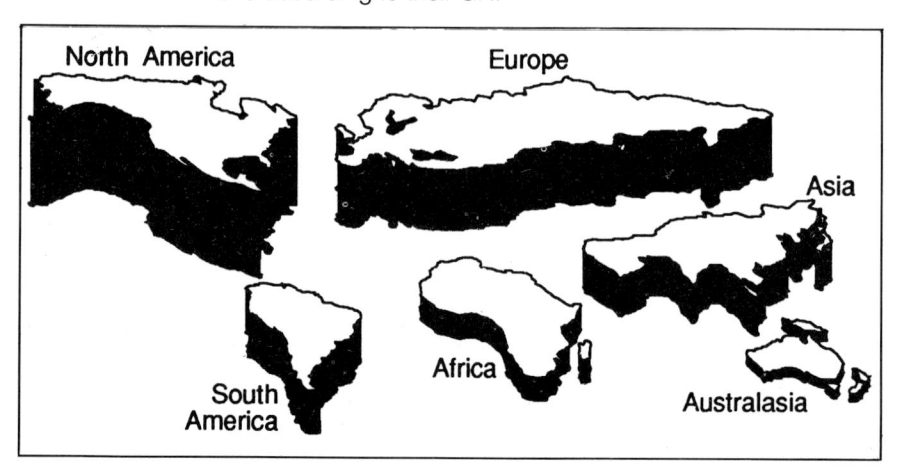

Source: *World Bank Atlas*.

UNEQUAL SHARES

In Britain we have more resources than people in other parts of the world.

1 Read how Peter Johnson describes his life in Soweto, South Africa.

> Our house is made of corrugated iron, beaten tin cans and pieces of wood. When it rains it leaks. We have no proper toilet or running water. We use a tap in the street. My father hurt his leg in an accident and can't walk properly. He earns a little money cleaning cars for rich people but we never really have enough to eat. When I'm not at school I spend most of my time playing with friends in the street. When I grow up I'd like to leave Soweto and see some other parts of the world but I don't think I'll have the chance.'

2 Write a few sentences about your own home, life needs and leisure facilities.

3 Look at the list of resources. Colour one circle on the first diagram for each of the things which Peter has. Do the same for yourself on the second diagram, then make a drawing of yourself in the space in the middle.

RESOURCES

Clean water supply	Opportunities for travel
Proper toilets	Friends
Clothing	Medicine and health care
Weatherproof housing	Leisure facilities
Schools	A balanced diet

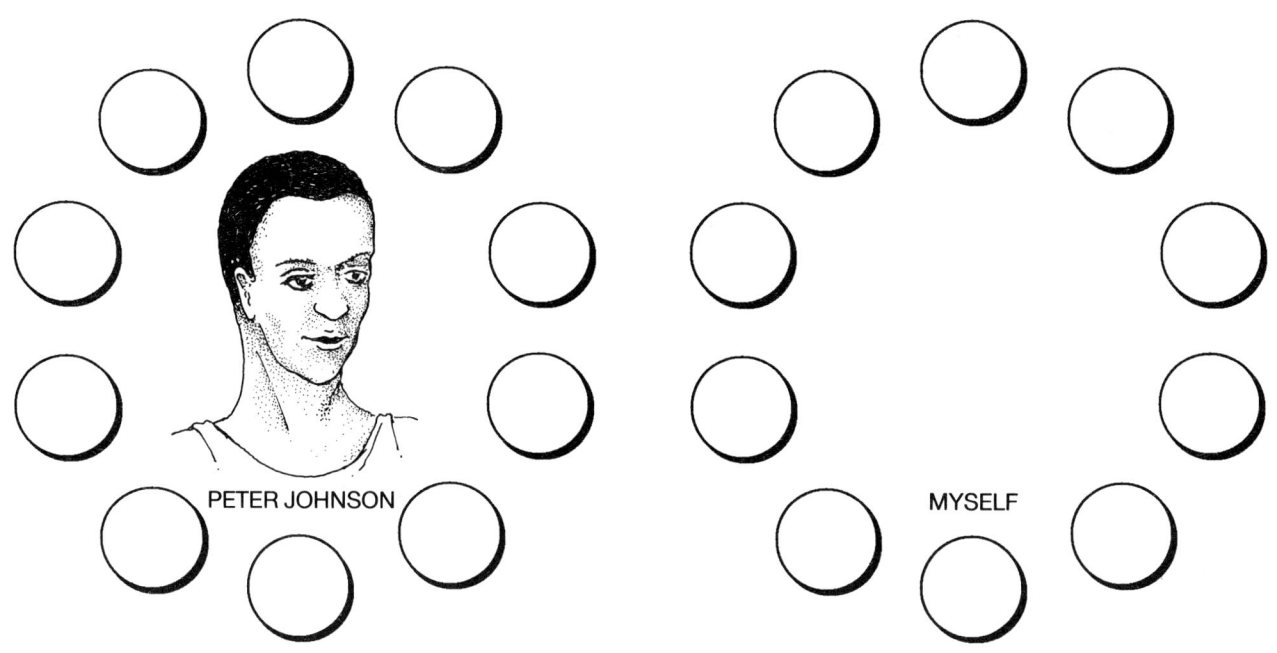

PETER JOHNSON   MYSELF

4 Compare the resources available to Peter and yourself by colouring the correct number of squares in the diagram below.

PETER JOHNSON										
MYSELF										

→ Deprived → Privileged →

© Scoffham, Bridge, Jewson, 1991. Macmillan. *Enquirybase Geography, Book 3*   ACTIVITY 26

STRUCTURE ACTIVITY 27

POINTS OF VIEW

People respond to the same place in different ways. Their perception of a city, for example, may depend on their economic well-being, political beliefs or social standing. If we are to understand the world in which we live, we therefore need to investigate human perceptions. This is an approach which has received considerable attention in recent years. Humanistic geographers, in particular, argue that we should look beyond facts and figures to the mainsprings of behaviour. In this activity pupils compare how three different people, a tourist, an executive and and a Hispanic immigrant, might view New York. Not only does this highlight the fact that we interpret our surroundings, it also illustrates that great inequalities can be found in a comparatively small area.

Skills
Reading for information
Tabulating data
Writing reports
Considering different needs/points of view

Attitudes and values
Consider the question of bias and perspective. How are we influenced by racial and cultural stereotypes? Do we assume that Britain is more advanced than other less wealthy countries? What do we mean when we say somewhere is remote? Would it make any difference if south, rather than north, was at the top of the world map?

Lesson preparation
1 It helps if pupils know something about New York before they begin the exercise.
2 A certain amount of inference is needed in completing the table. You might build up a profile for the tourist, executive and immigrant. It is particularly important that the pupils realise that the immigrant will have arrived in New York to escape desperate poverty in Mexico or Central America.
3 Pupils will need a variety of information about their locality to complete Question 3. The local plan and city guides will be useful sources.

Local enquiry/homework
1 Make a 'conceptual' map of your locality. Mark on it the things you think are important. See how it compares with maps drawn by other people. What are the similarities and differences?
2 What different images do people have of your area? Draw up a list of all the things which contribute to its character. Ask a range of people, preferably of different ages, to select the five things they think most important. Write a short report on your findings.

Extended investigation
What are the factors promoting urban growth in different parts of the world? Find out which towns are expected to have a population of over a million by the end of the century. Plot them on a map and try to explain the pattern. What are the advantages and disadvantages of city life? Make a collection of pictures or photographs of scenes you particularly like. Do they appeal equally to other members of the class? How do you think a blind or disabled person might respond to them?

Problem
Make a map of places in your locality which make you feel depressed, excited, uneasy and relaxed.

Source: *Contemporary Issues in Geography and Education*, Vol. 1, No. 2

POINTS OF VIEW

Our point of view affects how we respond to our environment.

1 The table below contains some facts and figures about New York. Decide if each one would be of interest mainly to a tourist, an executive or a Hispanic immigrant. Show your answer by colouring the correct circle. You can use more than one if you wish.

2 Using this information, write a short description saying why you think each of the three people decided to come to New York.

Facts and Figures about New York	tourist	executive	immigrant
It was founded as a trading centre in the seventeenth century by a Dutch trading company.	○	○	○
In 1950 the population was about 12 million. This could be nearly double by the year 2000.	○	○	○
About a quarter of the population was born abroad.	○	○	○
Although some wages are low, there are plenty of work opportunities.	○	○	○
There are some 1500 galleries and 92 museums.	○	○	○
Some of the world's largest banks such as Citibank and Chase Manhattan, are based in New York.	○	○	○
There may be as many as 400,000 illegal workers in the New York area.	○	○	○
Although the temperature drops well below freezing in winter, the summers are hot and humid.	○	○	○
Every year there are half a million robberies and assaults. Murders average over 1500.	○	○	○
In 1986 AIDS was the main cause of death in women aged 25–9. Poor Hispanics were especially vulnerable.	○	○	○
New York has better air communications than almost any other city.	○	○	○
Many tourists are attracted by the Empire State Building, Statue of Liberty and Broadway theatres.	○	○	○
The New York stock exchange is the most important in the world.	○	○	○

tourist

executive

Hispanic immigrant

3 Make a similar collection of facts and figures about your own town and neighbourhood. Try to find out about the number of people who live there, the facilities, how people earn their living, current social problems and any changes that are happening. Classify the information in a similar way and write descriptions from the points of view of (a) a tourist, (b) an executive, (c) a foreign immigrant.

© Scoffham, Bridge, Jewson, 1991. Macmillan. *Enquirybase Geography, Book 3*

PROCESS ACTIVITY 28

DIFFERENT DIETS

In the past, people produced the food they needed directly from the soil. Although this was a healthy way of eating, rural diet was, until recently, subject to massive seasonal variations. There was no adequate method of storing surpluses and in poor years there was always a danger of famine. Nowadays the situation has changed. Food production has become an extremely sophisticated process. Preservatives and packaging mean that food can be stored for long periods, and pesticides and fertilisers have raised levels of production. In the developed world shortages are no longer a danger. What is in doubt is the nutritional value of processed food. This activity introduces the question by getting pupils to compare their diet with that of a child in rural China.

Skills
Making a list
Tabulating data
Making comparisons
Using a pie chart

Attitudes and values
You might discuss whether it is morally right for countries like Britain to import food from countries where people are starving. More specifically, you could consider the advantages and disadvantages of processed food. What do pupils think about irradiation? Are any of them vegetarians? If so, why?

Lesson preparation
1 Explain that the questions in the two sections of the table are mutually exclusive. Thus, if the food is 'unwrapped or loosely bagged' it cannot also be 'sealed, tinned or packed'.
2 There are one hundred questions in all, fifty of which should receive a positive answer. To obtain a percentage pupils will need to multiply by two.
3 Discussion may be helpful when completing the survey on Fu Huang's diet.

Local enquiry/homework
1 Carry out a survey of food for sale at a local supermarket. Decide if each item is processed or unprocessed.
2 Make a detailed study of ten different processed foods. List the different colourings, flavourings and preservatives used in each one. Find out the meaning of any 'E' numbers mentioned on the label.
3 Arrange a visit to a local food processing plant, bakery or dairy. Make a study of the different processes by which the food is produced.

Extended investigation
Find out about the production and consumption of food in different parts of the world. Which countries have a food surplus, which ones need to import food? Compare the average calorie intake in Britain and other countries. In which areas is the problem of malnutrition particularly acute? How could it be alleviated? Make a study of famines, both at the present moment and in the past. Trace the history of food processing and preservation (refrigeration). Make an assessment of the advantages and disadvantages.

Problem
What do you eat that is produced within a radius of fifty kilometres?

In Britain the average person consumes over three kilograms of food additives a year.

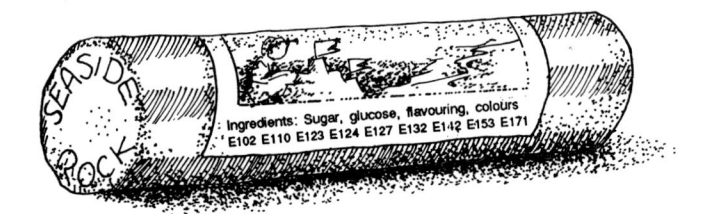

Source: *New Society and Statesman*, 12.9.86

DIFFERENT DIETS

Much of the food that we eat has been manufactured or processed.

1. On the table below write down ten food items which form part of your normal diet.

2. Colour the code boxes above each group of questions.

3. Think about each food item in turn and put a tick or a cross against each question.

4. Add up the totals of ticks.

	green						red					
	UNPROCESSED						PROCESSED					
FOOD ITEMS	unwrapped or loosely bagged?	comes direct from grower?	contains one ingredient only?	grown locally?	free from colour or preservative?	TOTAL	sealed, tinned or packed?	comes from a shop?	contains several ingredients?	comes from outside the area?	contains colour or preservative?	TOTAL

5. Look at a list of food eaten by Fu Huang in Hubei province, China. Complete the table in the same way as before.

rice from fields (grown by village)												
potatoes from family vegetable plot												
plums from local orchard												
eggs from chickens (kept by family)												
oil from rape seed (grown by village)												
salt from market (comes from elsewhere)												
meat from pigs (kept by family)												
barley from fields (grown by village)												
tomatoes from family vegetable plot												
fish from river												

6. Colour the pie charts to show how much of the food that you and Fu Huang eat is processed and how much is unprocessed. (Multiply your totals by 2 to obtain a percentage.)

7. What main differences do you notice?

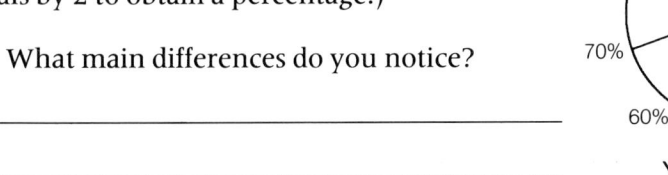

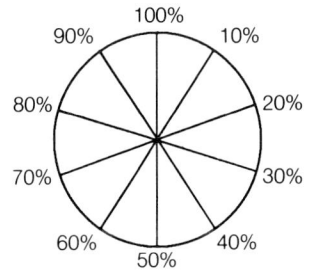

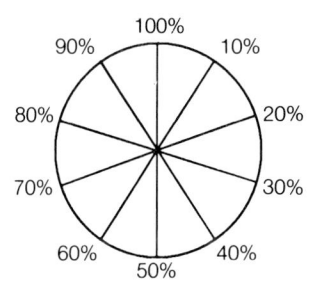

_____

_____

YOUR DIET   FU HUANG'S DIET

© Scoffham, Bridge, Jewson, 1991. Macmillan. *Enquirybase Geography, Book 3*   ACTIVITY 28

PROCESS ACTIVITY 29

MEDICAL SERVICES

Poverty is the single most important cause of ill health. Poor diet, polluted drinking water, inadequate housing and sheer ignorance trap millions of people around the world in a cycle of debilitation and disease. Every week a quarter of a million children under five die in developing countries. Most of these deaths could be prevented by low-cost health measures. In this activity pupils compare their own use of medical services with a comparable group of children in Bulawayo, Zimbabwe. This illustrates the enormous differences in provision and shows why it is a humanitarian imperative for those with power to tackle the problem.

Skills
Conducting a questionnaire
Making comparisons
Constructing a line graph
Drawing conclusions

Attitudes and values
Should everybody have the right to basic health care, regardless of their ability to pay? Do poor countries really have the money to pay for a basic health service? Will better care encourage unwelcome population growth? You could discuss these questions with reference to a few selected case studies.

Lesson preparation
1 Introduce the activity by discussing the different types of medical services available in the locality.
2 Some pupils will need to delve into their memories to answer the questions in the survey. In cases of doubt they should opt for the most probable answer.
3 See that the pupils understand how to complete a line graph. The scores must be recorded on the correct vertical axis. A system of dots, dashes or crosses will help to distinguish the two sets of figures.

Local enquiry/homework
1 Make a list of the different medical services that you use – doctor, dentist, chemist, optician, clinic, hospital and so on. Mark the location of each one on a map of your area and construct a spoke diagram showing how far they are from your home.
2 Find out about the history of medical services in your area. When was the local hospital first built? Have there been any major epidemics or health scares? Include articles from the local paper, field sketches and information from books in your answer.

Extended investigation
Trace the story of different infectious diseases such as smallpox, influenza and malaria. Draw maps showing the areas worst affected. How is the disease controlled? Is vaccination effective? Find out about medical services in different parts of the world such as the flying doctor in Australia and the barefoot doctors of India. Compare life expectancies in different countries. Is there any evidence that some places are naturally healthier than others? Make a case study of the British National Health Service. How does it compare with the private health-care schemes used in the USA?

Problem
Design a leaflet of health facilities for a newcomer to your area.

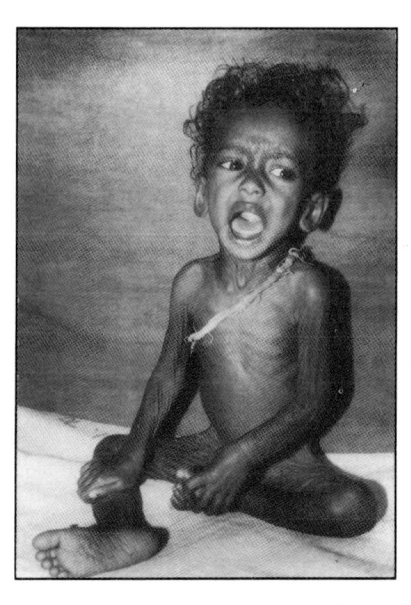

Source: *New Internationalist*, November 1986

There are 14 million child deaths each year. At least 10 million could be easily prevented.

Disease	Number of deaths
Diarrhoeal diseases	5 million
Malaria	1 million
Measles	1.9 million
Respiratory infection	2.9 million
Tetanus	0.8 million
Other	2.4 million

Source: *The State of the World's Children*, Oxford University Press, 1988

MEDICAL SERVICES

Medical services vary greatly in different parts of the world.

1 Complete the survey by putting a tick or a cross to show if you have ever used any of the services listed.

2 Collect information from eleven other pupils.

NAME OF PUPIL	DOCTOR			DENTIST	OPTICIAN	NURSE			HOSPITAL SERVICES				
	tablets for an infection?	a hearing test?	a school medical?	fillings in your teeth?	an eyesight test?	tetanus vaccination?	measles vaccination?	first aid?	X-rays?	stitches?	plaster (for broken bone)?	blood test?	an operation?

3 Add up the totals. TOTAL

4 Look at the results from the same survey taken at Katanga High School in Bulawayo, Zimbabwe.

6	3	6	4	1	5	6	10	2	3	1	0	0

5 Show the results of both surveys by drawing line graphs in the space opposite. Label them.

6 Colour the graphs using different colours for significant areas.

7 Write down any services for which the score differs by six or more points. In which place is the service better?

SERVICE	BETTER PROVISION

© Scoffham, Bridge, Jewson, 1991. Macmillan. *Enquirybase Geography, Book 3*

ACTIVITY 29

PROCESS ACTIVITY 30

LEARNING FOR LIFE

One of the factors that distinguishes human beings from other animals is the length of childhood and the enormous amount of time and attention lavished on the young. Education, in its broadest sense, is crucial for the developing child and one of the chief ways of transferring human knowledge and experience from one generation to the next. In this activity pupils analyse the source of their own skills and abilities. This shows the importance of schooling in their lives. Using the adult literacy rate as an indicator, they then consider the situation in other parts of the world and plot the results on a map to show the division beween North and South. The question of the quality and relevance of education in both Britain and other parts of the world is also implied.

Skills
Making a survey
Making comparisons
Identifying regions
Constructing a shaded area plan/map

Attitudes and values
Is schooling important? Are illiterate people stupid? Are there other and better ways of organising learning? What skills are most likely to give people opportunities? You might consider these questions from the perspective of a developing country as well as from personal experience.

Lesson preparation
1 Pupils must answer the survey honestly and not exaggerate their abilities.
2 Check that the pupils understand the term 'illiterate' and that they can identify the continents on the map.
3 Pupils must code all areas of the map and not just the main land masses.
4 Question 7 is designed to introduce the idea of North and South and should lead to discussion.
5 The world map uses the Eckert IV projection which shows area accurately but with less distortion of shape than in the Peter's projection.

Local enquiry/homework
1 Make a list of twenty-five things you know or have learnt to do. Decide if you have learnt each one (a) from an adult, (b) from a friend, (c) from the media, (d) taught yourself. Which of these four sources have had most importance in your life?
2 Working from a telephone directory, identify the different educational services available in your area – schools, libraries, museums, colleges and so on. Mark each one on a map and show which section of the population it caters for, using a code.

Extended investigation
Find out about the education system in different parts of the world. Is there any evidence that the provision is better for boys or girls? Are there any differences between urban and rural areas? Make a sample study of child labour in a selected industry. Trace the development of schools and schooling in Britain. To what extent is education geared to the needs of industry and commerce? Is there a connection between ignorance, poverty and disease?

Problem
Make a charter of the rights you think children ought to have.

Different attitudes to education

I left school when I was eight. The studies just would not go in. It just seemed all very difficult. I left because I had to start work. We didn't have money at home.
Luis Tacheco (14) COLOMBIA.

I am trying very much at present to improve my way of living by having education to obtain a well dignified job.
Phillip Samoel (15) KENYA.

To me school is just something you go to to get it over with. I never learn anything. It bores me. If my head was set on something, I could do it. But most school things – forget it.
Josie Keller (14) CANADA.

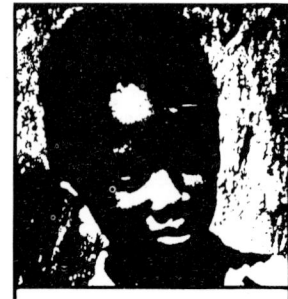

If you don't go to school it kind of mash up your life.
Kingsley Stewart (11) JAMAICA

Source: *Teaching Development Issues 6*, Oxfam

LEARNING FOR LIFE

Many of the things we learn are taught to us at school.

1 Write your name on the table below and colour a box for each skill you possess.

2 Using your knowledge of your parents and relations, make a similar survey of an adult.

3 Complete the survey by colouring the ticks in the columns for Mr Fernando and his son, Jose, who is fifteen years old. They both work as peasant farmers in north-east Brazil.

4 Add up the totals.

NAME OF PERSON	EARLY YEARS		PRIMARY SCHOOL YEARS				SECONDARY SCHOOL YEARS					ADULT YEARS					TOTAL
	talking	catching a ball	reading	writing	counting to 100	learning about own country	carrying out science experiments	learning a second language	learning algebra	growing vegetables	playing an instrument	herding animals	cooking food	repairing machinery	driving a car	earning a living	
Jose Fernando	✓	✓	✓	✓	✓						✓		✓		✓		✓
Mr Fernando	✓	✓			✓					✓	✓	✓	✓	✓	✓	✓	

Who has the most school skills?
_____

Who has acquired the most skills on their own?
_____

5 The table below shows the percentage of illiterate adults in different areas of the world. Colour the code box red if 5 per cent or more are illiterate, green if the figure is less than 5 per cent.

AREA	Europe	North America	Latin America	Africa	North Asia	East Asia	South Asia	Australia
PERCENTAGE OF ILLITERATES	2%	2%	17%	54%	2%	29%	56%	2%
CODE								

6 Write the names of the areas on the map then colour code each one using the code.

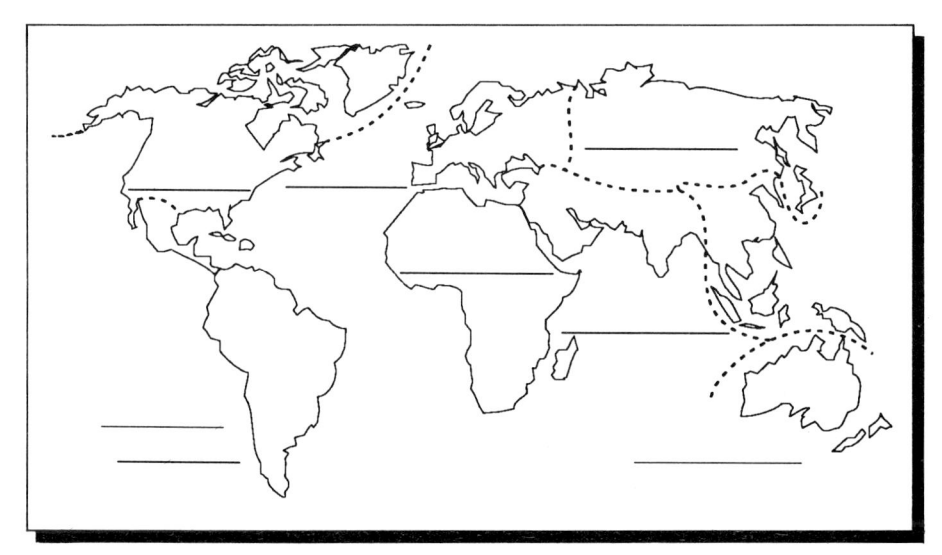

North America	
Europe	
North Asia	
East Asia	
South Asia	
Africa	
Latin America	
Australia	

N ↑ ↓ S

7 Colour code the North/South diagram using the same system.

© Scoffham, Bridge, Jewson, 1991. Macmillan. *Enquirybase Geography, Book 3*   ACTIVITY 30

SYSTEM ACTIVITY 31

POLLUTION

Attitudes to the environment have changed enormously since the early 1960s. Ecologists were the first people to draw attention to the importance of environmental issues and the way they relate to each other. Books like *Silent Spring* and *Limits to Growth* spelled out the scale of the problem. The same message was graphically illustrated by a series of disasters like the Torrey Canyon, Bhopal and Chernobyl. Now the Green Movement commands considerable support throughout the developed world. The question is whether it will be possible to avert even worse disasters in the future. In this activity pupils investigate a range of pollution problems from cigarette smoke to acid rain. By analysing the results diagramatically, they identify which ones seem most serious and think about how they might be solved.

Skills
Making a list
Tabulating data
Transferring information to a diagram
Making value judgements

Attitudes and values
Human history reaches back thousands of years but pollution seems to be a relatively new problem. You might discuss if this is really true, and why pollution is so difficult to solve. Try to think of the success stories as well as the problems which are threatening us nowadays.

Lesson preparation
1 Pollution is a complex concept. By way of preparation, discuss what it actually means with the class. Is it possible to reach a working definition which covers natural events like volcanic explosions, as well as human activity?
2 In Question 2 pupils can colour only one circle in the table to show the area affected.
3 Explain how to use the pollution diagram. It might help to plot a few examples by way of demonstration.

Local enquiry/homework
1 Make a survey of pollution problems in and around your school. Decide if each one could be solved by improved personal behaviour or whether it requires expenditure by the Local Education Authority.
2 Visit a number of different streets in your neighbourhood and decide how each one rates in terms of (a) litter, (b) noise, (c) air purity, (d) building quality, (e) plants and trees. Score each feature on a scale from 1 (very good) to 5 (very bad). Then add up the totals to arrive at a pollution index.

Extended investigation
Identify areas of the world where the environment is particularly at risk, such as the North Sea, Soviet Central Asia and the Amazon Basin. What is the nature of the threat, and how could it be solved? Make a case study of a major pollution disaster and the damage that it caused. Consider how natural vegetation and wildlife are affected by pollution problems. What international organisations have been set up to deal with pollution?

Problem
What two new laws, passed by the British Parliament, would be most effective in reducing pollution worldwide?

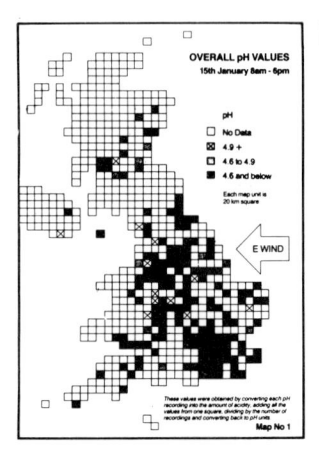

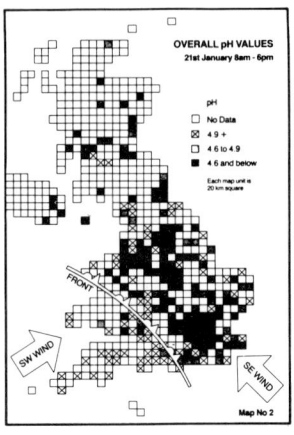

In 1985 WATCH, the national club for young people who care about the environment, conducted a survey into acid rain. The results showed how dirty European air is swept away by frontal systems from the Atlantic.

Source: *Acid Drops WATCH Trust for Environmental Education*

POLLUTION
Some types of pollution do more damage and are harder to solve than others.

1 Look at the list of different types of pollution. Add four more examples.

TYPE OF POLLUTION	AREA AFFECTED			DURATION								ACTION			
	local	regional	national	5 minutes	1 hour	1 day	1 week	1 month	1 year	5 years	25 years	individual	community	national	international
1 Aircraft noise	○	○	○												
2 Acid rain	○	○	○												
3 Leak from chemical factory	○	○	○												
4 Oiled beaches	○	○	○												
5 Smoke from a cigarette	○	○	○												
6 Ugly signs, wires and adverts	○	○	○												
7 Litter in the street	○	○	○												
8 Damage to the ozone layer (aerosols)	○	○	○												
9	○	○	○												
10	○	○	○												
11	○	○	○												
12	○	○	○												

2 Think about each example in turn, then:
(a) shade a circle to show if the area affected is local, regional or national
(b) draw a thick line to show the length of time it is likely to last
(c) colour a box to show how the problem might be solved. Use blue for individual action, green for community, yellow for national and red for international action.

3 Show the different types of pollution on the diagram opposite, using numbered dots. Begin by selecting one of the three rings (local, regional or national). Then choose the correct quadrant (up to 1 hour, up to 1 week, up to 1 year, 5 years or more). Mark the dot anywhere you like in the space selected.

4 From your survey, which types of pollution seem most serious?

_____
_____

5 Which type of action do you think is most likely to be effective in solving pollution problems? Why?

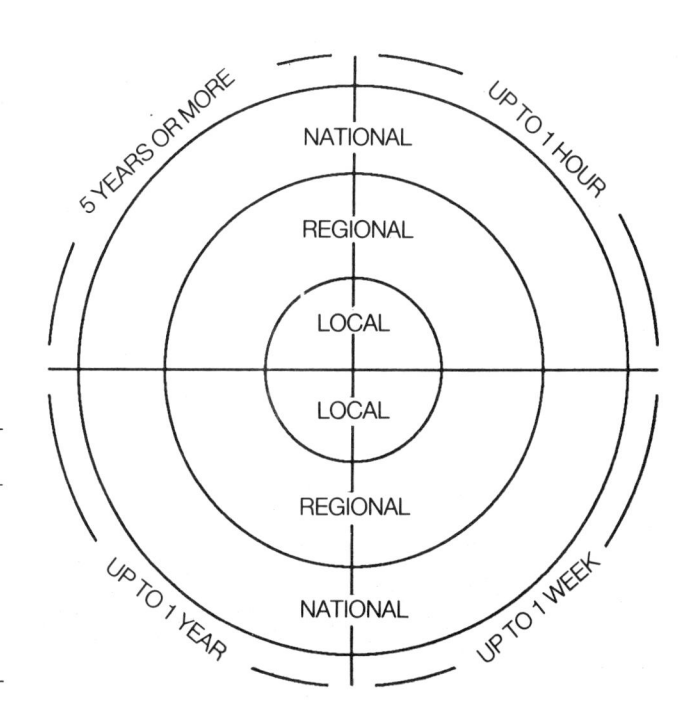

_____
_____

SYSTEM ACTIVITY 32

CONSERVATION

It is easy to be overwhelmed by the scale of environmental problems. Predictions of impending disasters do not always mobilise public support. They sometimes instil a sense of helplessness. Yet in some small way we are responsible for the care of our environment, and on a personal level we have a duty not to compromise our integrity. In addition, there can be little doubt that concerted action can achieve considerable success, as the consumer boycotts organised by Friends of the Earth have demonstrated. In this activity pupils conduct a questionnaire to discover how three people respond to environmental issues. The questions cover a range of topics including the recycling of materials, the conservation of fuel and other resources, and the work of pressure groups.

Skills
Conducting a questionnaire
Making comparisons
Using a scoring key
Writing a report

Attitudes and values
When there are five thousand million people in the world, how can our individual actions affect the problem of pollution? Why should we bother to behave responsibly? Is there any reason why we should not behave exactly as we want?

Lesson preparation
1. Introduce the activity by considering the meaning of conservation and how it differs from preservation.
2. When answering the questionnaire pupils must say what they *actually* do and not what they think they ought to do.
3. It is interesting if the questionnaire covers people of different ages and backgrounds. Pupils could complete it as part of a homework exercise.

Local enquiry/homework
1. Using a directory or by making enquiries at the local library, make a list of environmental groups or societies in your locality. Write a sentence describing the aims of each one. Then conduct a survey to see if ten random adults are aware of them.
2. Make up a prospectus for a school 'Conservation Club'. Design an eye-catching logo and say what you might do at your meetings during the first term.

Extended investigation
Make a study of an industry or form of agriculture which is sustainable in ecological terms. Find out how some creatures, such as grebes and kestrels, which were under threat in Britain have now made a comeback. Consider the measures being taken to preserve historic buildings and towns. How does the Government help to encourage conservation schemes? Which is more effective, legislation or economic incentives? If you had the choice, which areas in your locality would you preserve, and why?

Problem
Compile a directory or leaflet to help reduce pollution from households in your area.

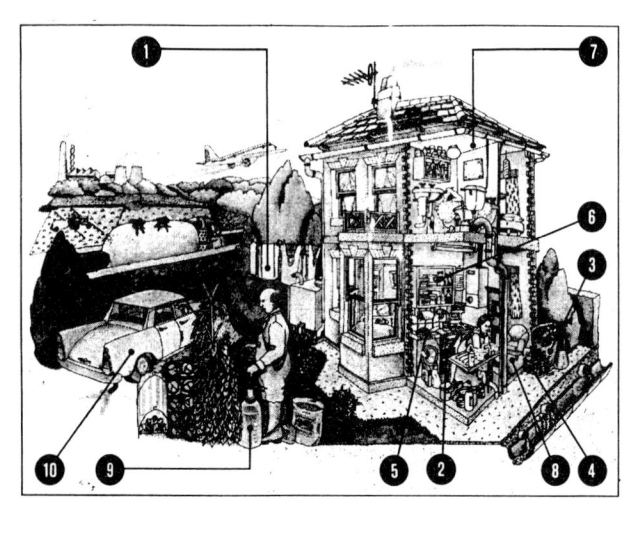

Ten ways to reduce indoor pollution

1. Use less detergent.
2. Buy free-range eggs.
3. Recycle kitchen waste.
4. Take bottles to a bottle bank.
5. Buy organically grown vegetables.
6. Say no to overpackaging.
7. Insulate your home.
8. Keep to natural fibres for clothes and carpets.
9. Ban biocides in home and garden.
10. Adopt low-pollution driving.

Source: *Observer*, April 1987

CONSERVATION

Our actions can either help to solve pollution problems or make them worse.

1 For each question below, colour one box to show your most likely action.

		Person 1	Person 2	Person 3	POINTS
(a) After drinking a glass bottle of orangeade, do you	(i) throw the bottle away in a bin?				
	(ii) leave it on a street corner?				
	(iii) take it to a bottle bank?				
(b) If you had an old bicycle you no longer needed, would you	(i) dump it somewhere?				
	(ii) advertise it in the paper?				
	(iii) take it to the rubbish dump?				
(c) When visiting a friend 3km away, would you rather	(i) walk or cycle?				
	(ii) go by bus?				
	(iii) go by car?				
(d) After peeling the potatoes for dinner, would you	(i) put the peelings in the dustbin?				
	(ii) feed them to the pet rabbit?				
	(iii) put them in a bucket for compost?				
(e) When you have finished reading your comics, do you	(i) give them to jumble?				
	(ii) throw them away?				
	(iii) burn them?				
(f) When you get cold watching TV, do you	(i) put on more clothes?				
	(ii) turn up the fire?				
	(iii) sit closer to the heater?				
(g) When you have finished with old clothes, do you	(i) see they are turned into rag?				
	(ii) throw them away?				
	(iii) give them to someone else to wear?				
(h) Do you	(i) give money for famine relief when asked?				
	(ii) think other people should fend for themselves?				
	(iii) belong to a conservation pressure group?				

2 Repeat the survey with two other people. If possible choose an adult so you get a good comparison.

3 Using this chart, copy down the points for each question.

a(iii) 1	b(iii) 2	c(iii) 3	d(iii) 1	e(iii) 3	f(iii) 3	g(iii) 1	h(iii) 1
a(ii) 3	b(ii) 1	c(ii) 1	d(ii) 2	e(ii) 1	f(ii) 2	g(ii) 3	h(ii) 3
a(i) 2	b(i) 3	c(i) 1	d(i) 3	e(i) 1	f(i) 1	g(i) 1	h(i) 2

4 Write each person's total in the box below.

Person 1	Person 2	Person 3

8-13 points	You are interested in the environment and think carefully about the effects of your actions.
14-18 points	You are reasonably aware of your surroundings but sometimes do thoughtless things.
10-24 points	Many of the things you do cause pollution. You need to think much more carefully about your actions.

5 Find out what the totals mean from the descriptions opposite.

6 Write a short report about how your personal lifestyle could be altered so that you become more conservation conscious.

© Scoffham, Bridge, Jewson, 1991. Macmillan. *Enquirybase Geography, Book 3*   ACTIVITY 32

SYSTEM ACTIVITY 33

DERELICT LAND

Estimates of the amount of derelict land in Britain vary, depending on the definition used, from a minimum of one per cent to a maximum of an area the size of Devon and Cornwall. Factors such as pollution, planning blight, historical accident and changes in agriculture and industry explain how dereliction develops. Given the political and economic will, nearly all the land could be brought back into production. In this activity pupils make a case study of a Kentish village. Working from an annotated map they decide how five different sites might be restored. This paves the way for a fieldwork exercise in the immediate locality.

Skills
Reading a plan/map
Using a colour code
Making choices
Drawing conclusions

Attitudes and values
Does it matter if there are patches of derelict land? What are the benefits? What are the disadvantages? If you were the Secretary of State for the Environment, what action would you be taking to deal with the problem?

Lesson preparation
1 By way of preparation, discuss different ways of defining derelict land. Consider the following formula: 'Land so damaged by industrial or other development that it is incapable of beneficial use without treatment'.
2 A certain amount of inference may be necessary in colour coding the different sites.
3 The suggestions in Question 4 are meant only as a prompt. Some pupils will want to think more imaginatively. It may also help if they identify Shepherdswell on a map of Kent so as to place it in a regional context.

Local enquiry/homework
1 Make a survey of derelict land in the area around your home or school. Mark each site on an annotated map, as in the activity sheet.
2 Investigate the ecology of one of the sites in your area. Identify the plants which are growing there and make careful sketches of one or two of them. List any birds and creatures that you find. Devise a food web for the whole habitat.

Extended investigation
Consider the problem of waste disposal, especially toxic and radioactive materials. What are the dangers? Which parts of the world (both land and sea) are most at risk? How do economic forces create derelict land? Make a sample study of the degradation of the Amazon rainforests. Identify different examples in Britain such as quarries, spoil heaps and old aerodromes. Find out about sites which have been reclaimed. What are they used for, and was it expensive to bring them back into use?

Problem
Invent a method for the disposal of radioactive waste.

How do the plants in your own town match this list of the twenty most common wild city flowers?

Bittersweet, *Solanum dulcamara*
Burdock, *Arctium minus*
Canadian fleabane, *Conyza canadensis*
Charlock, *Sinapis arvensis*
Coltsfoot, *Tussilago farfara*
Creeping thistle, *Cirsium arvense*
Curled dock, *Rumex crispus*
Dandelion, *Taraxacum officinale*
Fat hen, *Chenopodium album*
Golden rod, *Solidago canadensis*
Greater plantain, *Plantago major*
Knotgrass, *Polygonum aviculare*
Meadow grass, *Poa annua*
Mugwort, *Artemisia vulgaris*
Red clover, *Trifolium pratense*
Rosebay willowherb, *Epilobium angustifolium*
Rye-grass, *Lolium perenne*
Shepherd's-purse, *Capsella bursa-pastoris*
Smooth sowthistle, *Sonchus oleraceus*
Yarrow, *Achillea millefolium*

Source: *Street Flowers*, Richard Mabey Kestrel 1976

DERELICT LAND

Patches of productive land are left unused for a variety of reasons.

1 Look at the notebook below. On the left there is a map of Shepherdswell in Kent. On the right are the notes which Lucy Thompson made when she did a survey of derelict land.

MAP	SITE	DESCRIPTION	REMARKS
(map of Shepherdswell or Sibertswold showing sites 1–5)	Site 1	Embankment on disused railway	Overgrown but used as an exercise area for dogs
	Site 2	Small area of ancient downland	Used as a field for grazing horses
	Site 3	Derelict orchard	Very neglected. Adjoins small lorry depot
	Site 4	Small patch of empty land between houses	Used as a play area by children
	Site 5	Corner of farmyard	Used for dumping soil

2 Choose a colour for each of the code boxes in the key opposite.

3 Using these colours, complete the code boxes in the table below.

DESCRIPTION	CODE
polluted land that needs reclaiming	
small, awkward spaces	
areas waiting for a planning decision	
land left over from past activities	

SITE	CODE	POSSIBLE USE	BEST CHOICE
1		(1)	
		(2)	
2		(1)	
		(2)	
3		(1)	
		(2)	
4		(1)	
		(2)	
5		(1)	
		(2)	

4 Think about each site in turn. Using the suggestions opposite, list two possible uses. Remember, you can always leave the area unchanged if you wish.

5 Now decide how you think the site could best be used. Give reasons for your choice.

nature reserve	allotments	garage
car park	houses	village hall
farmland	warehouses	village pond
children's play area	grazing horses	new road
toilets	new woodland	shop

© Scoffham, Bridge, Jewson, 1991. Macmillan. *Enquirybase Geography, Book 3*

INTERACTION ACTIVITY 34

WHAT MATTERS MOST?

Politicians at all levels decide on the importance of different issues and allocate resources accordingly. In making decisions they are guided by a range of factors such as their own political beliefs, the influence of pressure groups, notions of social justice and financial considerations. In this activity pupils conduct a questionnaire on the problems affecting their school and the action being taken to deal with them. By plotting the results on a scatter diagram they then discover which ones are most pressing.

Skills
Establishing rank order
Using a scoring key
Drawing a scatter graph
Drawing conclusions

Attitudes and values
Problems do not always affect all sections of the community equally. Is it more important to deal with something which affects a few people seriously or something which affects a lot of people in a more minor way? Is there any point to this type of equation anyway?

Lesson preparation
1 Pupils could enter themselves on the table first before interviewing other people.
2 The same people have to be interviewed in Questions 1 and 2. The two tables can therefore be completed at the same time.
3 Some schools may not be affected by the problems in the list. It follows that nothing needs to be done and that the situation is satisfactory.
4 Check that the pupils understand how to complete the scatter graph at the bottom of the sheet.

Local enquiry/homework
1 Carry out a similar study in your locality or town. Make a list of problems, interview six people to discover if they are satisfied with what is being done, and show the results on a scatter graph. From this evidence do you think your local Council is tackling the right problems?
2 Draw an annotated map of problems in your neighbourhood. Devise symbols for each problem and try to divide them into different categories – urgent, average and minor.

Extended investigation
What problems are the most pressing in the world today? Make a list using the following headings – food, health, pollution, economics, resources, natural disasters, conflicts. How was each one caused? Can you suggest a solution? Make a study of the Brandt or Brundtland Report. Compile a scrapbook of newspaper articles on world problems. What international agencies have been set up to tackle these problems? Do you think they are effective?

Problem
On what issues should the local Council hold a referendum?

Survey of problems that most concern young people aged ten to seventeen.

Source: *The Guardian*, January 1987

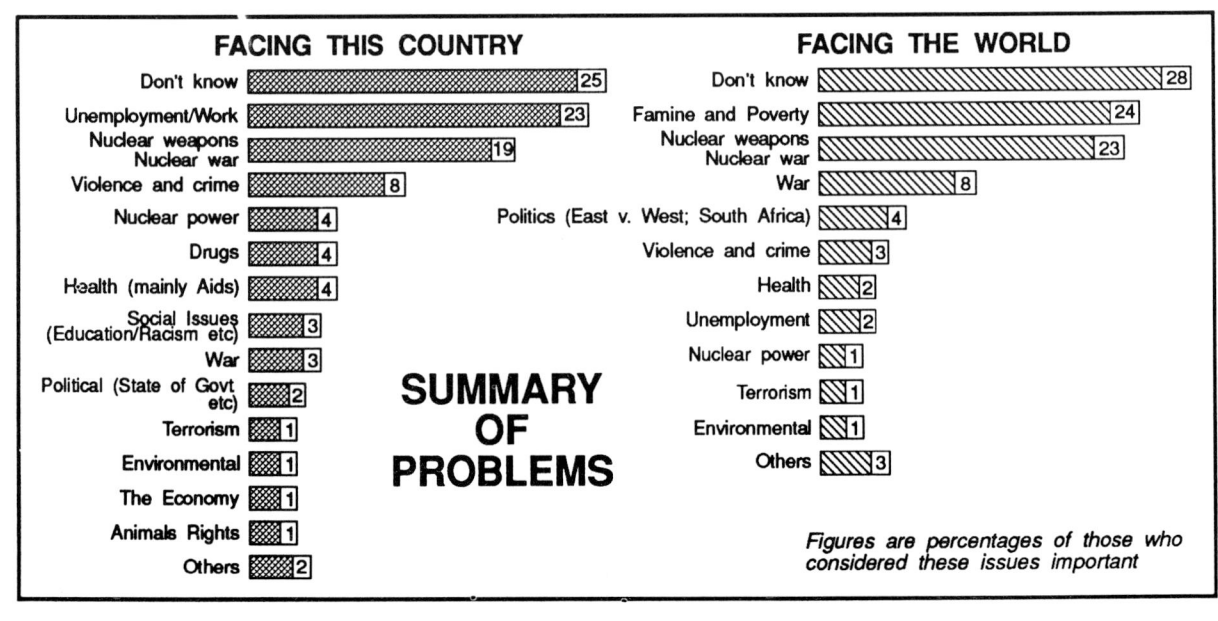

SUMMARY OF PROBLEMS

FACING THIS COUNTRY
- Don't know: 25
- Unemployment/Work: 23
- Nuclear weapons/Nuclear war: 19
- Violence and crime: 8
- Nuclear power: 4
- Drugs: 4
- Health (mainly Aids): 4
- Social Issues (Education/Racism etc): 3
- War: 3
- Political (State of Govt etc): 2
- Terrorism: 1
- Environmental: 1
- The Economy: 1
- Animals Rights: 1
- Others: 2

FACING THE WORLD
- Don't know: 28
- Famine and Poverty: 24
- Nuclear weapons/Nuclear war: 23
- War: 8
- Politics (East v. West; South Africa): 4
- Violence and crime: 3
- Health: 2
- Unemployment: 2
- Nuclear power: 1
- Terrorism: 1
- Environmental: 1
- Others: 3

Figures are percentages of those who considered these issues important

WHAT MATTERS MOST?

Some problems are more urgent than others.

1 Look at the list of problems in the table opposite. Ask six pupils to place them in order of importance. Write 1 next to the most important problem, 2 next to the next most important and so on. Then add up the totals.

2 Now ask the same six pupils if they are satisfied with what is being done to deal with the problem. For each question circle the score which best describes their opinion, using the following system.

2 points   satisfied
1 point   OK
0 points   not satisfied

Add up the totals as before.

PROBLEM	PERSON						TOTAL
	A	B	C	D	E	F	
1 Classes too large?							
2 Not enough practical work?							
3 Old books and equipment?							
4 Poor school dinners?							
5 Split school site?							
6 Repetitive lessons?							
7 Poor sports facilities?							
8 Too much homework?							

PROBLEM	PERSON						TOTAL
	A	B	C	D	E	F	
1 Classes too large?	2 1 0	2 1 0	2 1 0	2 1 0	2 1 0	2 1 0	
2 Not enough practical work?	2 1 0	2 1 0	2 1 0	2 1 0	2 1 0	2 1 0	
3 Old books and equipment?	2 1 0	2 1 0	2 1 0	2 1 0	2 1 0	2 1 0	
4 Poor school dinners?	2 1 0	2 1 0	2 1 0	2 1 0	2 1 0	2 1 0	
5 Split school site?	2 1 0	2 1 0	2 1 0	2 1 0	2 1 0	2 1 0	
6 Repetitive lessons?	2 1 0	2 1 0	2 1 0	2 1 0	2 1 0	2 1 0	
7 Poor sports facilities?	2 1 0	2 1 0	2 1 0	2 1 0	2 1 0	2 1 0	
8 Too much homework?	2 1 0	2 1 0	2 1 0	2 1 0	2 1 0	2 1 0	

3 Mark each problem with a numbered dot on the graph opposite. Plot the level of importance against the horizontal axis and the level of satisfaction against the vertical axis.

4 From your graph, which problems are most pressing?

_____

_____

Which problems are least pressing?

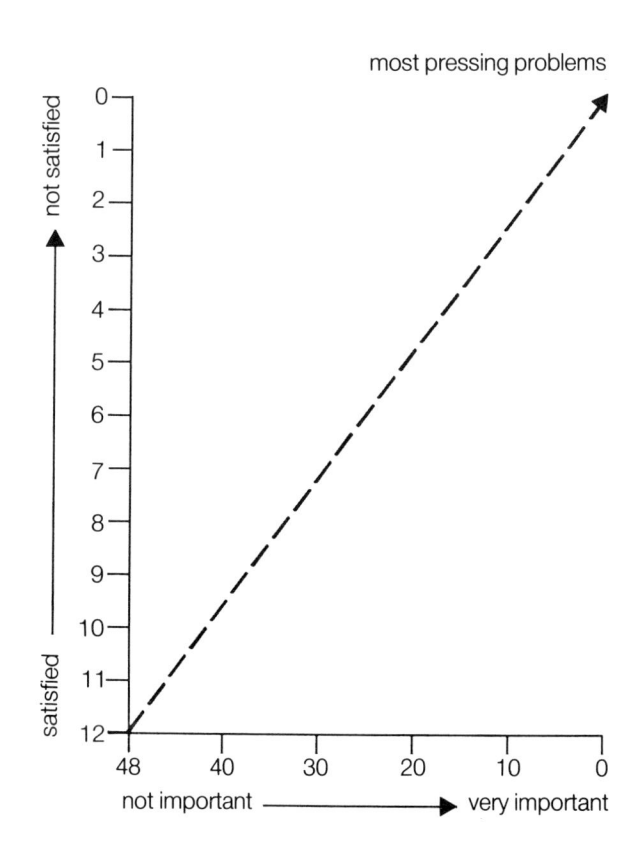

_____

_____

© Scoffham, Bridge, Jewson, 1991. Macmillan. *Enquirybase Geography, Book 3*

ACTIVITY 34

INTERACTION ACTIVITY 35

CHOICES

Previous activities have shown that it is vital that people adopt positive attitudes towards the environment (Pollution problems and Conservation). Although individual decisions may seem insignificant, when multiplied nationally they become crucially important. This activity considers two major questions, namely living standards and family size. Pupils conduct a survey to discover preferences in their own class. They then discover from a flow diagram the possible consequences if lots of other people made the same choice. A final section shows how Britain compares with other countries on these same questions of population and economic growth.

Skills
Conducting a questionnaire
Reading a flow chart
Making value judgements
Transferring information to a diagram

Attitudes and values
The activity raises general questions about progress and the quality of life. It also suggests that present choices affect future generations. In China, for example, the Government has passed laws redistributing the land and restricting family size. Why don't other countries try to stabilise living conditions in a similar way?

Lesson preparation
1 The survey in Question 1 can involve any number of pupils from a small group to an entire class. What matters is that the results should show a definite preference or choice.
2 The figures for population and economic growth are for the period 1980–5 and are taken from *The State of the World's Children* (UNICEF). More up-to-date figures could be substituted when they become available.
3 Check that the pupils understand how to plot the figures on the graph.

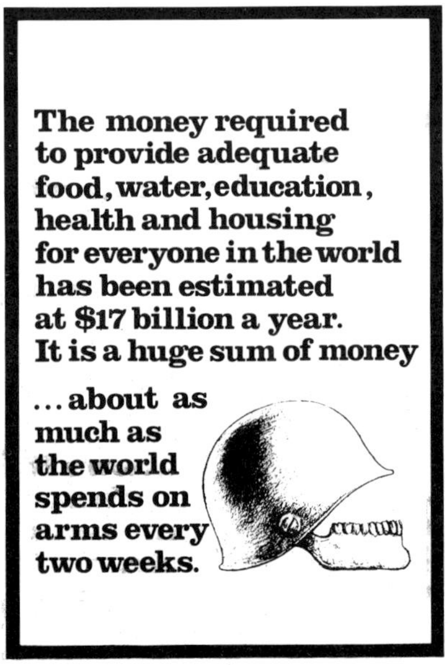

Local enquiry/homework
1 Make a study of conservation and action groups in your locality. What are their aims? How do they try to achieve them? Draw a map showing the location of nearby sites and reserves, e.g. National Trust, RSPB and so on.
2 Consider your own scope for making choices that will affect the environment (a) in your home, (b) in your street, (c) in your locality. What things can you affect directly, which things can you influence, and which are the ones over which you have no power? Make a survey.

Extended investigation
Make a case study of one of the countries mentioned in the activity sheet. Find out especially about the way it has changed over the last thirty years and try to trace the consequences of particular government policies and actions. Make a list of people who have influenced history through their individual action. Say what each one did. What single discoveries have most influenced (a) the modern world, (b) twentieth-century life?

Problem
Imagine you are given a large sum of money. How would you use it in an influential way?

The money required to provide adequate food, water, education, health and housing for everyone in the world has been estimated at $17 billion a year. It is a huge sum of money ...about as much as the world spends on arms every two weeks.

Source: *New Internationalist*

CHOICES

The choices we make nowadays will affect our well-being in the future.

1 Ask the pupils in your class what they want when they grow up, using the questionnaire below. Each question has two parts and they must answer 'yes' to one of them. Record their replies by colouring a box in the correct column. Then add up the totals and say what decision your class has reached.

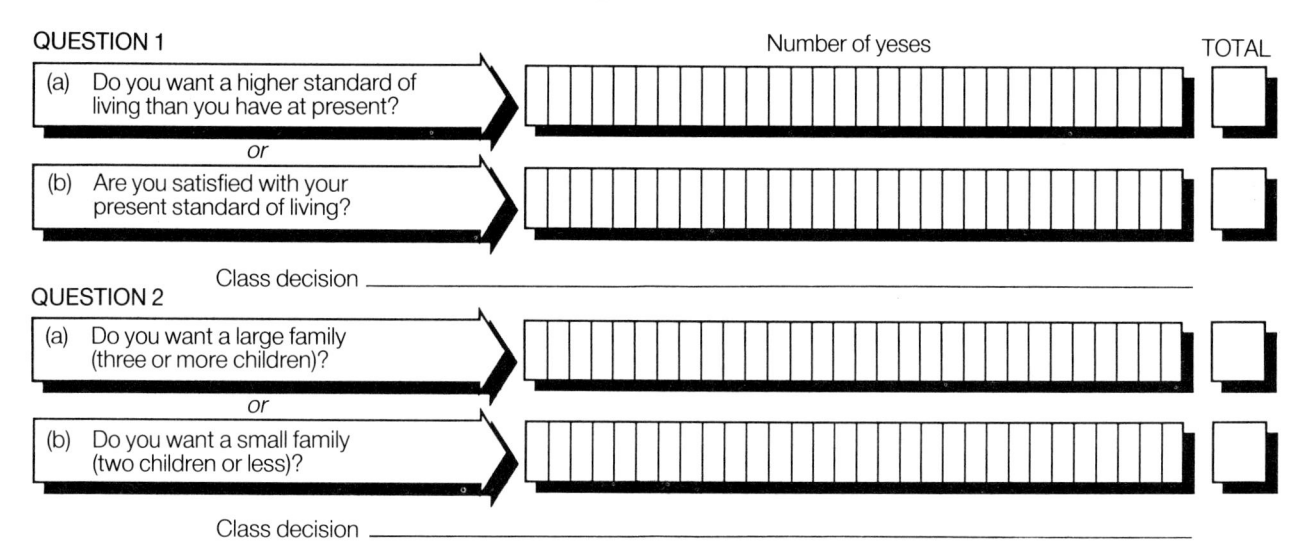

QUESTION 1 — Number of yeses — TOTAL
(a) Do you want a higher standard of living than you have at present?
or
(b) Are you satisfied with your present standard of living?

Class decision _____

QUESTION 2
(a) Do you want a large family (three or more children)?
or
(b) Do you want a small family (two children or less)?

Class decision _____

2 Follow the flow chart to see what could happen if a lot of other people made the same decision as your own class. Draw a coloured line to show the trend.

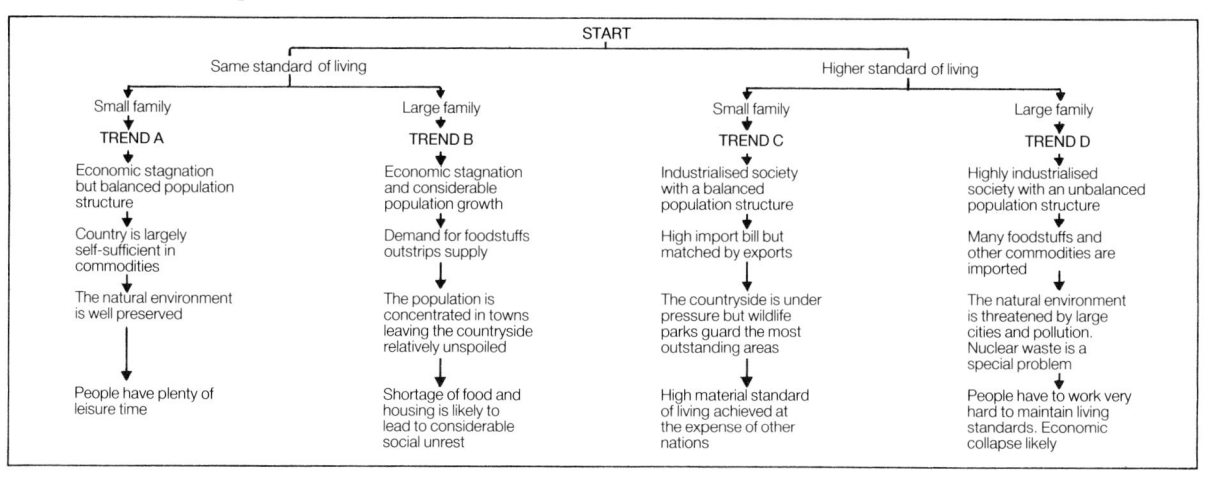

START

Same standard of living
- Small family → TREND A
- Large family → TREND B

Higher standard of living
- Small family → TREND C
- Large family → TREND D

TREND A
- Economic stagnation but balanced population structure
- Country is largely self-sufficient in commodities
- The natural environment is well preserved
- People have plenty of leisure time

TREND B
- Economic stagnation and considerable population growth
- Demand for foodstuffs outstrips supply
- The population is concentrated in towns leaving the countryside relatively unspoiled
- Shortage of food and housing is likely to lead to considerable social unrest

TREND C
- Industrialised society with a balanced population structure
- High import bill but matched by exports
- The countryside is under pressure but wildlife parks guard the most outstanding areas
- High material standard of living achieved at the expense of other nations

TREND D
- Highly industrialised society with an unbalanced population structure
- Many foodstuffs and other commodities are imported
- The natural environment is threatened by large cities and pollution. Nuclear waste is a special problem
- People have to work very hard to maintain living standards. Economic collapse likely

3 Are there any problems with the trend your class has chosen? What are they?

4 Compare the countries listed in the table by plotting numbered dots on the graph opposite.

COUNTRY	Annual population growth	Annual economic growth	TREND CODE
1 Britain	0.1%	2.1%	
2 USA	1.0%	1.4%	
3 Japan	0.7%	3.5%	
4 Bangladesh	2.6%	0.9%	
5 India	2.6%	3.1%	
6 Hong Kong	1.4%	4.4%	

Graph: population growth (y-axis, 0 to 3%) vs economic growth (x-axis, 0 to 6%), divided into four quadrants: B (top-left), D (top-right), A (bottom-left), C (bottom-right).

5 Say which trend each country is following at the moment.

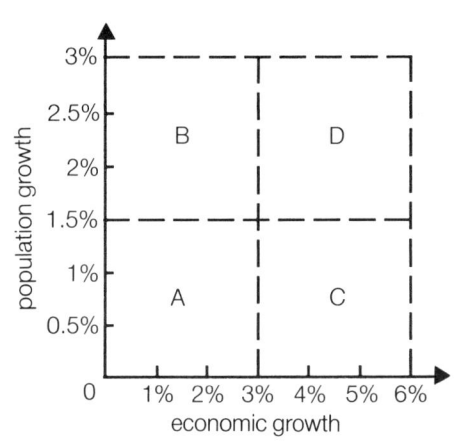

6 Which country best matches your class choice?

© Scoffham, Bridge, Jewson, 1991. Macmillan. *Enquirybase Geography, Book 3*   ACTIVITY 35

INTERACTION ACTIVITY 36

TOMORROW'S WORLD

Many of the previous activities have shown that the world is beset by problems. These range from social injustice and inequality to environmental issues and international conflict. This activity takes a more idealistic stance. It invites pupils to identify the three problems they would most like to solve and considers what action would be most effective in solving them. This highlights the importance of global issues and stresses the role of political action in dealing with the challenges that confront us as we enter the twenty first century.

Skills
Making choices
Tabulating data
Making value judgements
Writing a report

Attitudes and values
All human beings have an idealistic streak. How does this vision affect the way we interpet the contemporary world? What organisations and institutions encourage idealism? Is there really any place for idealism nowadays, or should we be concentrating on the problems which surround us?

Lesson preparation
1 Discuss the choices listed in the table. Get the pupils to add a few suggestions of their own by way of introduction.
2 In Question 3 pupils can tick as many different forms of action as they think would be effective.
3 Question 4 is designed to promote discussion and to prepare the ground for the reports they are asked to write in Question 5.

Local enquiry/homework
1 On what issues would you campaign if you were standing as an independent candidate for the local Council? Devise a leaflet describing your policies and the things you think are important. Try to persuade others to vote for your proposals in a class simulation.
2 Describe how you think your locality might change in the next twenty years, using the following headings: population, housing, transport, industry, recreation. Use maps and diagrams in your answer.

Extended investigation
Make a study of political systems in different parts of the world. Have they evolved or have they been imposed? Draw maps showing the way different countries are governed. Look at the map of Europe. How has it changed over the last two hundred years? Show the different political boundaries. Find out about the history of the European Community. What effect does the European Parliament have on life in Britain? Think about the work of the United Nations. Why was it set up? Has it been successful in dealing with conflicts? What are the prospects for the future?

Problem
Make a profile of the sort of person you want to be as an adult.

ATTRACTIVE, DETACHED RESIDENCE

Believed to be unique, this magnificent dwelling has been sadly neglected in recent years. Some outstanding features have been lost. However it still offers an exceptional home to those prepared to maintain it with care.

The earth has evolved over a period of 4600 million years. A nuclear war could render it unhabitable in a few seconds.

Source: Friends of the Earth Poster

TOMORROW'S WORLD

Political decisions have a crucial influence on the world we live in.

1 Look at the list of choices in the table below. Ask eight different people to vote for the three things they would most like to happen. Record their replies by putting an X in the correct column. Write any extra ideas in the empty spaces.

2 Add up the totals.

3 Think about each choice in turn. Put a tick against the different forms of action you think would be most effective.

CHOICES	VOTES									ACTION			
	person 1	person 2	person 3	person 4	person 5	person 6	person 7	person 8	TOTAL	All countries must agree	Scientific discovery	People must change	More money needed
1 Everyone would have a job													
2 Nuclear weapons would be abolished													
3 There would be no more wars													
4 There would be a cure for AIDS and other diseases													
5 There would be no more famines													
6 Animals would be killed only for food													
7 Everyone would have clean water													
8 Torture would be abolished													
9 There would be no more crimes of violence													
10 There would be no oppression of minorities													
11 All pollution problems would be solved													
12 The rainforests would be preserved													
									TOTAL				

4 Colour this column if you think political action would make a significant difference.

5 Use the information from the survey to write a brief report about
(a) what would make tomorrow's world a better place, and
(b) how this might be achieved.

6 How well does the UK match up to these requirements?

© Scoffham, Bridge, Jewson, 1991. Macmillan. *Enquirybase Geography, Book 3*   ACTIVITY 36

© Stephen Scoffham, Colin Bridge, Terry Jewson 1991

All rights reserved. No reproduction, copy or transmission of this publication may be made without written permission except under the terms set out below.

This publication is copyright, but teachers are free to reproduce the activity sheets by any method without fee or prior permission, provided that the number of copies made does not exceed the amount required in their school. For copying in any other circumstances (e.g. by an external resource centre) prior written permission must be obtained from the publishers and a fee may be payable.

Any person who does any unauthorised act in relation to this publication may be liable to criminal prosecution and civil claims for damages.

First published 1991

Published by
MACMILLAN EDUCATION LTD
Houndmills, Basingstoke, Hampshire RG21 2XS
and London
Companies and representatives throughout the world

Designed and produced by Hart McLeod, Cambridge
Typeset by Goodfellow & Egan, Cambridge

Printed in Great Britain by
Martin's of Berwick Ltd.

British Library Cataloguing in Publication Data
Scoffham, Stephen
　World issues. — (Enquirybase geography; book 3).
　1. Geography
　I. Title II. Bridge, Colin III. Jewson, Terry
　IV. Series
　910

ISBN 0–333–45925–3

Acknowledgements

The authors and publishers wish to thank the following who have kindly given permission for the use of copyright material.

Centre for Global Education for figure from *World Studies Journal*, 7, 1, 1988; HarperCollins Publishers Ltd. for figure from *Systematic Geography* by Brian Knapp, Allen & Unwin, 1986; and cartoon from *Architecture for Beginners* by Louis Hellman, Writers and Readers, 1986; The Controller of Her Majesty's Stationery Office for material from *OPCS Spotlight No. 4*; Metheun London for cartoon from *The Effluent Society* by Norman Thelwell, 1971; New Statesman Society for material from September, 1986, and July, 1988, issues; Open University Press for material from *The Third World Atlas*, 1983; Punch Publications Ltd. for cartoon by David Hawker, *Punch*, Jan. 1983; Watch Trust for Environmental Education Ltd.

We are grateful to the following for permission to reproduce photographs: Aerofilms; Hunting Areofilms; World Health Organisation.

Every effort has been made to trace all copyright holders, but if any have been inadvertently overlooked the publishers will be pleased to make the necessary arrangement at the first opportunity.